Green Chemistry Applications

Edited by Murat Eyvaz and Ebubekir Yüksel

Published in London, United Kingdom

IntechOpen

Supporting open minds since 2005

Green Chemistry Applications
http://dx.doi.org/10.5772/intechopen.77499
Edited by Murat Eyvaz and Ebubekir Yüksel

Contributors
Shu-Peng Ho, Prateek Khare, Ravi Shankar, Ratnesh Kumar Patel, Sachin Pandurang Shirame, Ragunath Bhosale, Rania H. Taha, Joel Martínez, René Miranda-Ruvalcaba, Nilce Ortiz

First published in London, United Kingdom, 2019 by IntechOpen
IntechOpen is the global imprint of INTECHOPEN LIMITED, registered in England and Wales, registration number: 11086078, The Shard, 25th floor, 32 London Bridge Street
London, SE19SG – United Kingdom
Printed in Croatia

British Library Cataloguing-in-Publication Data
A catalogue record for this book is available from the British Library

Additional hard and PDF copies can be obtained from orders@intechopen.com

Green Chemistry Applications
Edited by Murat Eyvaz and Ebubekir Yüksel
p. cm.
Print ISBN 978-1-78984-123-7
Online ISBN 978-1-78984-124-4
eBook (PDF) ISBN 978-1-78984-404-7

For EU product safety concerns:
IN TECH d.o.o., Prolaz Marije Krucifikse Kozulić 3, 51000 Rijeka, Croatia, info@intechopen.com or visit our website at intechopen.com.

Meet the editors

Dr. Murat Eyvaz is an assistant professor in the Environmental Engineering Department at Gebze Technical University (GTU), Turkey. He received his BSc in Environmental Engineering from Kocaeli University, Turkey, in 2004. He obtained his MSc and PhD in Environmental Engineering in 2013 from the Gebze Institute of Technology (former name of GTU). He completed his post-doctoral research at the National Research Center on Membrane Technologies in 2015. His research interests are water and wastewater treatment, electrochemical processes, filtration systems/membrane processes, and spectrophotometric and chromatographic analyses. Dr. Eyvaz has co-authored numerous journal articles and conference papers and has taken part in many national projects. He serves as an editor for 40 journals and a reviewer for more than 115 journals indexed in SCI, SCI-E, and other indexes. He has one patent and three patent applications on wastewater treatment systems.

Prof. Ebubekir Yüksel is a faculty member of the Environmental Engineering Department at Gebze Technical University (GTU), Turkey. He received his bachelor's degree in Civil Engineering from İstanbul Technical University in 1992. He obtained his MSc and PhD from İstanbul Technical University in 1995 and 2001, respectively. His research interests include applications in water and wastewater treatment facilities, electrochemical treatment process and filtration systems at lab. and pilot scale, watershed management, flood control, deep sea discharges, membrane processes, spectrophotometric analyses, chromatographic analyses, and geographic information systems. Professor Yüksel has co-authored numerous journal articles and conference papers and has taken part in many national projects. He has produced more than 25 peer-reviewed publications in journals indexed in SCI, SCI-E, and other indexes. He has two patents and three patent applications on energy and wastewater treatment systems.

Contents

Preface

With the emergence of the concept of green chemistry, which is defined as the use of a dozen principles that reduce or eliminate hazardous materials in the design, manufacture, and use of chemical products, it is understood that focusing on precautions to reduce or eliminate existing pollution sources is more effective than looking for a cleaning path after exiting. In parallel with increasing and diversifying needs today, with the help of technological developments, production has also increased and diversified. However, in addition to the technical and economic suitability of the processes leading up to the final product, the elimination or reduction of adverse effects that might occur to our health and the environment is now being taken into consideration.

This book describes the principles of green chemistry for clean production in light of the latest technological developments and increasing environmental awareness. As regards industrial applications, in order to minimize wastes at the beginning of production, the book presents research on the development of alternative raw materials, new catalysts for use in the production stage, non-volatile solvents, non-hazardous chemicals, and rapidly renewable and reusable products that are biodegradable in nature by using renewable energy. This book consists of six chapters covering topics such as synthesis and applications of nanomaterials for energy and environmental applications, climate process, alternative green energy sources, and removal of emerging pollutants from water.

Murat Eyvaz
Assistant Professor,
Department of Environmental Engineering,
Gebze Technical University,
Turkey

Ebubekir Yüksel
Professor,
Department of Environmental Engineering,
Gebze Technical University,
Turkey

Chapter 1

Recent Developments in *Nanocarbon-Polymer Composites* for Environmental and Energy Applications

Prateek Khare, Ratnesh Kumar Patel and Ravi Shankar

Abstract

Advancement in technology is brought up for both the industrialization and urbanization, which imitates the diverse pollutants as by-products to air-water ecosystem. Utilization of carbon is a very well-known age-old practice used for environmental remediation. In recent past years, a tremendous research inclination has been witnessed toward the sustainable development of nanocarbons specifically for environment remediation and energy applications. The new sustainable synthesis approached has been investigated for the development of sustainable nanocarbons. Tailoring the nanocarbon with polymer as a nanocomposite material has spotlighted a new class of environmentally benign bio-compatible materials. Herein, the present perspective is summarizing the basic-advanced research.

Keywords: nanocarbon, polymer composite, bio-polymer, environmental remediation

1. Introduction

Use of chemical products such as, conventional energy reserves, synthetic detergents, pharmaceuticals, dyes, fertilizers, polymers and composites, food additives, agrochemicals, etc., increases over period time across the developing and developed countries and simultaneously is very important for the development of any economy. The society is consuming such products at an exponential rate and thereby generating pollutant in terms of fuel combustion, waste generation in the process operation and waste disposal [1]. Till date, due to their maximum operational, and consumption of the conventional energy reserves including, fossil fuel, and coal are the prime factors of increasing the pollutions to the biosphere at such alarming rate. Moreover, their residue generates unwanted carbon like carbon soot interact ecosystem imparts their toxic nature and affects the ecology [2–4].

Developments of the green environmentally benign materials and its competent with technology advancement are very necessary and now have become a challenging task for the developing eco-friendly society in the biosphere. The concept of zero-waste production, waste prevention, and with use of efficient materials is the prime concerned for the industries, Therefore, for a healthy society, the role of green chemistry is significant, achieving these targets [5]. As it is a well-known mechanism that green chemistry offers a relatively lesser toxic synthesis approach, by reducing the harmful chemical substances in the designing the useful chemical products [6]. Nowadays, due to the unique properties of carbon-based nanomaterials like good electrical conductivity, ease in surface functionalization, high mechanical strength and good thermal stability of carbon-based fullerenes carbon nanotubes (CNTs), carbon quantum dots and graphene, attracted high interest toward research and used widely in many application purposes. Therefore, in the reported literatures, as wonder materials, carbon-nanomaterials have been used directly or modified for aforesaid applications [7]. Although, following complex techniques and expensive hydrocarbons or other specific hazardous source like laser are not easy to handle in different synthesis routes like chemical vapor deposition (CVD), plasma CVD, laser ablation used for the synthesizing the different carbon-based nanomaterials. There are some reported literatures in which an inexpensive and environmentally friendly approach is exploited for recovering the carbon based nanomaterials and experimentally particularly shown their applicability in remediation and sensing applications [8–14]. The synthesis of nanocomposites plays leading role in the current advanced applications purposes such as, energy storage, electronics parts, environmental remediation, biomedicine, etc. [15].

Present book chapter deals with production and potential applications of nanocarbon and nanocarbon-polymer composite materials, with special attention on the energy and environmental related sector and their significant role in enhancing the efficacy for the previously mentioned applications.

2. Classifications of carbon and carbon-based nanomaterials

Carbon have tendency to polymerize into the large molecular weight compounds with long chains due to its unique electronic construction and the smaller size in comparison of group IV that makes it capable of linking with other elements. As it contains four electrons in the valence layers, carbon can easily form covalent bonds with both metals and non-metals. Due to this property, carbon-based compounds can exist in diverse molecular forms, and same type of atoms can be arranged in the different shapes, with different properties known as allotropes. Graphite and diamond are two known natural occurring allotropes of carbon found in the ecology and formed other multiple allotropes of carbon from natural carbon sources for the different purposes. Furthermore, with the knowledge of top down and bottom up approaches of nanotechnology, a new family of carbon nanomaterials as wonder materials is introduced. In recent past years, the different types of carbon nanomaterials have been experimentally tested and successfully developed as advance materials in many engineering applications. In general, carbon nanomaterials exist in between 1 and 100 nm size range; therefore, the unique properties of carbon nanomaterials like good electrical, ionic conductivity, high mechanical and thermal stability at the nanoscale compared with other materials are exceptional and competent for many engineering applications [16]. Carbon nanomaterials are classified further based on their shape and geometrical structures, till dates different forms of carbon nanomaterials that are existing; horn-shaped as nanohorns, tube-shaped as carbon nanotube (CNT), ellipsoidal spherical shape carbon nanospheres (fullerenes), and zero-dimension dots exhibited quantum character as carbon quantum dots (CQDs). Because of excellent properties, carbon nanomaterials have now been extensively used synthesizing the nanocomposite materials that and used in many applications including microchannels, micro/nanoelectronics, textiles, paints, gas storage, composites, conductive plastics, displays, antifouling batteries with excellent durability during cycling-charge. The carbon quantum dots, as a newly emerging biocompatible material, have exhibited fluorescence properties and possess many other valuable excellent properties, such as high aqueous solubility, low cost, low toxicity, abundant surface functional groups and large active surface area for functionalization. Although, besides being used as bio compatible materials, such new class of (carbon quantum dots) nanomaterial are also applied as gas biosensors and heavy metal sensing applications [7, 17, 18]. In particular, the unique features of CQDs as up-converted photoluminescence (PL) behavior and photo-induced electron transferability of CQDs have proposed a new route for harvesting the sunlight from nonconventional reserve, to achieve efficient metal-free photocatalysts [19–21].

2.1 Fullerenes

Fullerene as a new class of carbon family had introduced by Kroto, Curl and Smalley [22] in 1985 and denoted as buckminsterfullerene (C60). It is an intermediate allotrope of carbon between graphite and diamond as it constitutes from the bunch of atomic Cn repeating unit ($n > 20$) which are collectively composed to form a spherical surface having a hollow core, or empty region of space inside the molecule. Carbon atoms are usually located on the surface of the sphere at the vertices of pentagons and hexagons and linked by forming sp2-hybridizing covalent bonds. Generally, C60 has two bond lengths, in which double bonds for 6:6 ring bonds are shorter than the 6:5 bonds. The important distinct characteristic ofC60 is that the pentagonal rings resulted in poor electron delocalization. As a result, C60 behaves like an electron deficient alkene, and reacts readily with electron rich species. C60 is the odorless, and nonvolatile black solid having density 1.65 g/cm^3, standard heat of formation 9.08 kcal/mol with boiling point of 800 K [23]. Fullerenes, due to its excellent properties like superconductivity, radical scavengers and excellent durability, are frequently applied in medicinal and electronics field, also in solar cell, fuel cell, cosmetic products (low order fullerene such as, C28, C26, etc.), biological/medical area, catalysis and other relevant fields [24–26]. Furthermore, fullerene can be easily modified to tailor properties for nanocomposites synthesis. [27].

2.2 Carbon nanotubes

CNTs are one of the types of allotrope among carbon-based nanomaterials that have excellent mechanical and electrical properties. CNT are light-weighted, high strength material as compared to steel at nondimensional scale and discovered first by a Japanese scientist S. Ijimain 1991. Nowadays, due to its extraordinary graphic nature and high specific surface area have attracted attention and applied as a pillar material in many engineering applications like battery, electrochemical, pollutant remediation, and used as fillers for nanocarbon-polymer composites [28–30]. CNT exist broadly in two different shapes; cylindrical shaped is formed by rolling of the graphene sheets; and possessed cap fullerene structure appeared in half shape. Based on their geometrical configuration as proficiently qualified is experienced by high voltage electron microscopy techniques. CNT are further classified into two types (i) single-walled carbon nanotube (SWCNT), and (ii) multiwalled nanotube (MWCNT), Carbon nanotubes are prepared in rolled sheets of very few single layer carbon atoms (graphene) form cylindrical molecules. CNT with diameter 1–3 nm and length of few micrometers while for MWCNT, graphene sheets having 0.34 nm of inter-space distance, are stacked like concentric layers in cylindrical form of diameter 5–40 nm. Zhang et al. has been described in a reported literature, which is approximately 550 mm in length [31]. The properties of CNT are basically dependent upon the diameter, size and morphology. There are several methods are reported for CNT preparation like arc discharge, laser ablation, chemical CVD, and plasma CVD. Among the listed methods, arc discharge was the first technique used for the preparation of CNT while laser-ablation method was used to prepare the SWCNT. In the chemical CVD method, small amount of metallic catalysts (Ni and Co) are used to catalyze the hydro-carbon as source at relatively lower temperature for the growth of graphitic surface. The high enhanced electrical property of SWCNTs is due to the presence of the chirality or hexagon orientation with respect to the tube axis, however on bulk scale its synthesis process is very complex and not easy to control the layers. In contrast to the SWCNT, due to the presence of multiple layers, MWCNT possess high mechanical and thermal satiability. Further, based on SWCNTs morphology, it is classified into three subgroups: (i) armchair morphology exhibiting high electrical conductivity than the copper, (ii) zigzag morphology has good semiconductor property and (iii) chiral morphology has semi-conductive property.

2.3 Graphene

Graphene is two dimensional, single-atom layer of carbon atoms which are sp2 hybridized and fixed in a rigid hexagonal lattice like a flat plane. Graphene is also a primitive building element of graphite, fullerene and CNT, Graphene was discovered in 2004 by Canadian physicist Wallace. It is an allotropic form of carbon with bond length of 0.142 nm between neighboring atoms of carbon and layer by layer of graphene is stacked with an interplanar spacing of 0.335 nm. The layers of graphene in graphite are bounded by Van der Waal forces [32, 33]. The unique physical properties of graphene, such as, thermal stability, mechanical rigidity and electrical conductivity are higher for few layers of graphene than of their three-dimensional materials. Also, graphene conducts high heat because of high thermal conductivity of graphene in comparison with available excellent heat conductors such as, silver and copper, and much better than graphite and diamond [7, 33].

2.4 Carbon quantum dots

A new class of with unique fluorescent property of carbon nanoparticles discovered accidentally Xu et al. in [34] during purification of SWCNTs. Later in 2006, Sun et al. had given a name of such fluorescent materials as carbon quantum dots (CQDs) particle of size found less than 10 nm. Till date, due to its fascinating property (harvesting optical light and imparting multicolor tuned emission) of CQDs offers a surprising potential material in fields of bio-imaging, photo-degradation and catalysis applications [35]. In last decade, various chemical precursors like citric acid, ammonium citrate, ethylene glycol, benzene, phenylenediamine, phytic acid, and thiourea, have been used for synthesizing CQDs. In order to minimize energy consideration, various synthetic methods, including hydrothermal, solvothermal, electrochemical, microwave assisted pyrolysis, ultrasonication, and chemical oxidation, etc., have been tested to produce the fluorescent CQDs. A number of review and research papers have been focused on the synthesis of such CQDs [36–39]. However, to date there has not been a very few reviews which explicitly focused on green synthesis routes is discussed in details for sensing and bio-imaging of applications [40].

Figure 1 describes the different types classification of synthesis routes used in developing the different types of nanocarbons.

2.5 Naturally occurring synthesis nanocarbon

Nanocarbons occur naturally, but not available at abundant scale; therefore, this approach is not very conventional to control the number of graphitic layers,

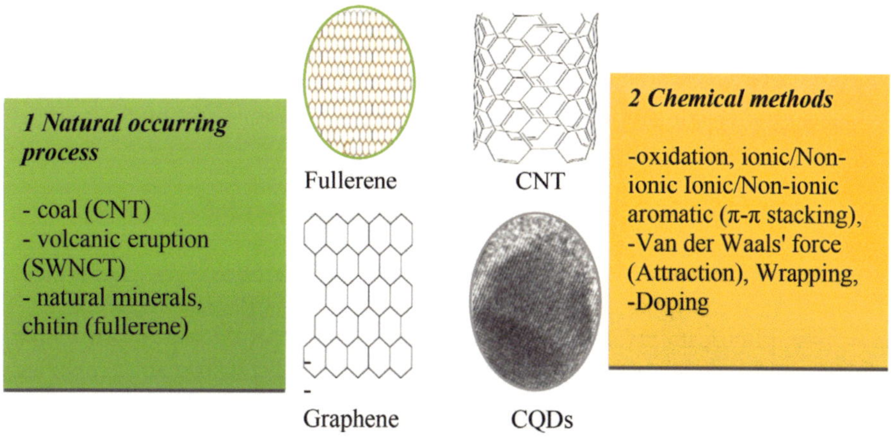

Figure 1.
Types of synthetic routes of different nanocarbons.

therefore their physical and chemical properties of nanocarbon may vary for engineering applications. There are different types of nanocarbon available from natural synthesis are reported in literature [41]. Velasco-Santos et al. described the existence of carbon nanotubes in the coal/petroleum mixture [42]. SWCNT can be synthesized by CVD, Su and Chen, 2007 and Mracek et al. 2011used metal oxide mixed volcanic lava as a substrate and catalyst [43, 44]. It was noted that process may provide indication for a probable creation of nanomaterials in natural conditions when the temperature rises extremely high, e.g., during volcano eruptions. Like CNT and SWCNT, fullerenes are also found in different ecological materials, for example in the natural mineral shungite from Karelia fullerene is found in low concentrations (2% w/w) [45] and also in meteorite samples of cosmic origin [46]. Chitin is one of the naturally occurring nanomaterials obtained from carbohydrate polymer. Synthesis of chitin nonmaterial considers the following factors such as, thermal dimensional stability, dispersibility, mechanical reinforcements, antibacterial activity etc. depending on the specific goals [47]. Natural nanomaterials can be obtained from polymer waste and its feasibility depends upon processing, recycling, transportation cost, etc. [48].

2.6 Chemically functionalized nanocarbons

Chemical functionalization process formed a huge distinct variety of carbon-based nanomaterials having different functionality, which are applied successfully in different sectors. As the surface functionalization means, carbon-based nanomaterials are added with other groups that ultimately changed its chemical and physical properties [49, 50]. There are many methods of functionalization available for carbon based nanomaterial, such as oxidation, ionic/non-ionic aliphatic aqueous (Hydrophobic), Ionic/Non-ionic aromatic (π-π stacking), Van der Waals' force (Attraction), Wrapping, doping, and direct deposition. The modification of surfaces is depending on the feasibility and degree of functionalization for the specific application. Therefore, in light of specific application, materials have been synthesized by following different mechanisms like non-covalent bonds, covalent bonds, electrostatic force, hydrogen bonds, Van-der Waals force etc. [51]. Some example of surface modification are as follows; Jiang et al. covered the CNT surface with active sulfate groups by using sodium dodecyl sulfate ($NaC_{12}H_{25}SO_4$) a common surfactant (anionic) for dispersing the sulfate groups [52]. In oxidation process CNT are exposed to mixture of acids via ultrasonic treatment method, during this process, carboxylic groups (–COOH) is attached with CNT surfaces. Oxidation of CNT is very important and creates oxygen carrying groups (–COOH and –OH), which makes CNT feasible for further functionalization without affecting their electrical and mechanical properties [7].

3. Potential applications of carbon-based nanomaterials

The outstanding physiochemical properties of nanocarbons have triggered interest, toward the applicability of nanocarbons, in multiple area including adsorption, photocatalyst, fertilizer, nanobiotechnology products, environmental materials, and renewable energy related application etc. De Volder et al. reported industrial scale production approximately up to several thousand tons of nanocarbon [17]. Properties like, high thermal stability and mechanical strength of nanocarbon make them suitable alternatively as fillers provides high aspect ratio for nanocomposites materials. Therefore, the prepared nanocomposites materials exhibited enhanced mechanical and other properties as compare to their starting primitive materials [53]. Nasibulin et al. reported the composite material based on cement matrix in which the addition of nanocarbon to original materials exhibited higher strength composite material [54]. Use of low weight, high strength materials in mechanical equipment improves the overall efficiency with low energy consumption. Regarding this use of nanocarbon based materials in energy generating machine such as, turbine as a lightweight and mechanically tough material is desirable [55, 56], and non-corrosive nature and insoluble nature of sp2 hybridized nanocarbon as fillers for marine turbines [57], various electronic applications [58], automotive industry [59], aviation [60], sport equipment allow to use strong and light weighted materials, which generates economical energy generation at optimal energy consumption. Fullerenes and composite based materials are frequently used in pharmaceutical industries [61]. Nowadays, graphene and its composite are frequently in high demands in various applications such as, electronics, solar cell, biochemical sensors [62]. Carbon based nanomaterials based composite materials having excellent properties like high tensile strength, flexibility and good electrical conductivity that make them more favorable for electronics applications [63]. Similarly, the graphene-carbon nanotube/polyvinyl alcohol based composite show high rigidity, strength, and ductility in comparison with conventional nanocarbon materials. Although there is a dramatically increase in the resistivity witnessed for graphene-carbon nanotube/polyvinyl alcohol-based composite, Therefore, such composite materials can be a suitable as a smart stretchable insulator devices using formed by combining the property of conductive nanocarbon materials with epoxy polymer [64].

3.1 Environmental applications of nanocarbon-polymer-based composites

Ecological balance is essential which is majorly affected by different pollutants. Therefore, it is difficult to check ecology balance simultaneously with the rapid growth of industrialization and civilization. In this regard, it is required to increase effectiveness by taking some corrective measures in pre-existing methods of controlling pollutions [65, 66]. In light of nanotechnology knowledge, advanced nanomaterial/composite materials can be developed that significantly enhanced the performance of different pre-existing treatment technologies [55, 57]. For example nanocarbon provide high specific surface area thus, carbon-based nanomaterial/composites provide high specific surface for the adsorption process [29, 30] oxidation process [67, 68], and electrochemical applications [69, 70]. Although, the nanocarbon is one of the most effective adsorbents used in the wastewater treatment because of its high surface area to weight ratio. But, it cannot be applied directly in flow or dynamics conditions [71]. Specific affinity of metal functionalized-nanocarbon based materials as adsorbent exhibited high adsorption capacity specific for different heavy metals (As, Fe, Pb, Cu), vitamin B12, nitrates, phosphates and show superior antibacterial activity than that of primitive activated carbon [72–77]. Moreover, after adsorption operation, specific adsorbate with nanocarbon based adsorbent can be separated by the mechanism of filtration and can be recycled [10]. But, high preparation cost of nanocarbon based materials and uncertainties regarding the leaching potential are the major challenges in the application of nanocarbon based materials. Silver coated carbon-based nanomaterials can be used as an antibacterial material and used in disinfection purposes, antimicrobial activity and preparation of biomedical devices [78]. Carbon nanotubes, activated with alcohol 1-octadecanol can also be used as better adsorbent for the absorption of microwaves because of its antibacterial characteristics. So, it can be used in the water purification technologies [79].

Moreover, nanocarbons are not only important in the pollution reduction in different ecological sections, but it has significant importance in the monitoring of pollution stages. Carbon-based nanomaterials and their composites can be used for the development of different and efficient sensors (biochemical) to detect very low concentrations of chemical compounds in different environments. For example, carbon nanotubes loaded with ZnO nanoparticles can be used for congo red dye reduction from aqueous solutions and showing that ZnO/MWCNTs is a promising, environmentally friendly and efficient adsorbent for wastewater treatment [80].

3.2 Energy applications of nanocarbon-polymer-based composites

Economy of any country is directly boom if they can produce and store energy particularly from renewable resource. In this viewpoint carbon-based nanomaterials and their composites can play an important role in the area of energy harvesting and energy storage. It is well known most of the nanocarbon composed of sp2 hybridization and possessed excellent properties such as, high pore size distribution, high surface area, with enhanced mechanical properties and improved electrical properties. In recent years, nanocarbon and nanocomposites are widely applied for developing energy storage and energy saving devices/instrument. Photovoltaic cells are commonly known as solar cell used as an alternative device for harvest renewable energy sources. Photovoltaic cells can be classified in two categories such as, thin films and crystalline silicon photovoltaic cells. Earlier, silicon, cadmium, copper-based compounds are used in the semiconductors used [81, 82] applied in energy storage devices. Nowadays thin-film group photovoltaic cells used platinum-based semiconductor for high band width/specific related applications. But, high cost and availability of platinum increases overall cost of an instrument. In this case carbon-based nanomaterials/composites can play an alternative role of platinum-based materials due to its superior properties [83]. Generally, nanomaterials prepared from graphene are used to enhance electron carriage and boost the efficiency of solar energy conversion [84]. Graphene-based materials can also be used in fuel cells and batteries due to its favorable properties.

Supercapacitors are used for energy storage devices applied in electric vehicles, hybrid electric vehicles, backup power cells, and portable electronic devices due to its advanced properties such as, high-power density, very short charging time, and high cycling stability. The main mechanism of storage of energy in super capacitors are pseudo capacitance and electrochemical double-layer capacitance. In pseudo capacitor, faradic reactions mechanism is responsible for charge transfer processes.

	SN	Nanomaterials	Method	Materials used	Conditions	Application	Ref
Environmental Application	1	Multiwall carbon nanotube	Oxidation, filtration, drying	H_2SO_4: HNO_3 (3:1)	Sonicated; 3 h; 40°C	Heavy metal removal (Cadmium reduction)	[67]
	2	Oxidized multiwall carbon nanotube	EDA, N - HATU, filtration, drying	NH_2 materials	Sonicated; 4 h; 40°C	Heavy metal removals (Cadmium and lead reduction)	[68]
	3	Multiwall carbon nanotube	Oxidation/reduction/pyrolysis	Ethanol/ferrocene/thiophene	Airflow, 400°C, argon flow, 600–900°C	ciprofloxacin reduction from aqueous solution	[90]
	4	Graphitic oxide	Oxidation, filtered; washed; dispersed	$NaNO_3$, H_2SO_4, $KMnO_4$, H_2O_2	Vigorous agitation 20–66°C, dilution: 98°C, 15 min	Wastewater treatment	[91]
	5	Graphite Oxide	Oxidation, sonicated; washed; dried	$Na_2S_2O_3$	Sonicated; 1 h, disproportionation; 30 min	Heavy metal removal (mercury Hg^{2+} reduction)	[92]

	SN	Nanomaterials	Method	Materials used	Conditions	Application	Ref
Energy Application	6	Graphite powder	Heating, drying, filtration, mixing	$K_2S_2O_8$, P_2O_5, H_2SO_4, $KMnO_4$, H_2O_2	Heating: 80°C, ice bath: 20–35°C, mixing: 35°C for 2 h	Fabricating of various microelectrical devices	[93]
	7	Graphene oxide, graphene nanoplatelets	Heating, drying, filtration, mixing	Polyethylene glycol, H_3PO_4, $KMnO_4$, H_2SO_4, HCl, H_2O_2	Mixing: 50-0 °C for 12 h, Drying: 50°C, sonicated; 1 h	To prepare higher thermal energy storage material	[94]
	8	Single wall carbon nanotube	Absorption, drying, Nitraticn	HNO_3, Ag/AgCl electrode, methylene blue	Nitration: 10 h, methylene blue doping: 3 h	Application in biofuel.	[95]
	9	carbon nanotubes	Melting/heating, sonicated	graphite nanoplatelets, phase change materials	Heating: 60°C, ultrasonicated at pulse velocity of 25 m/s for 20 min	To prepare enhanced thermal conductive material	[96]
	10	Porous carbon and hydrous RuO_2	Oxidation, absorption	Sulfuric acid		Supercapacitor for energy storage	[97]

Table 1.
Application of nanocarbon based materials in energy and environmental field.

Many metals/oxides/conducting polymers are good examples of the pseudo capacitance process. While, in electrochemical double-layer capacitance processes, charges are accumulated at the interface by the mechanism of adsorption/desorption process of electrolyte ions on a large surface area electrode materials. So, in this regards carbon-based nanomaterials can play an important role in the supercapacitor preparations [85]. Supercapacitors based on nanocarbon have many advantages over conventional (metal-based) supercapacitor, such as, high cycling stability, high power density and low energy density limits for their applications in batteries [86].

Excellent mechanical and electrical properties of nan0carbon-based materials (carbon nanotube) offer an exposed surface to functionalize and make them suitable for energy storage. But, it has some disadvantages such as, moderate capacitance due to low density of nanomaterials [87]. The lithium-ion battery is alternative type of energy storing substance, which holds energy as a chemical energy. It has many advantages over capacitors such as, high power density, and less greenhouse gas emissions possibilities [88]. Nanocarbon materials/composites are used in the lithium batteries because structure of the nanocarbon-based material usually express some common factors such as, the amount of lithium that is reversibly incorporated into the carbon lattice, the faradic losses during the first charge–discharge cycle, and the voltage profile during charging and discharging.

Carbon based nanomaterials such as, carbon nanotubes, activated carbons, and graphene based nanosheets are suitable for sustainable energy storage devices, because, carbon materials have many favorable properties such as, light weight, low cost, easy processability, adaptable porosity, and simplicity of chemical modification [89]. Generally, higher specific surface area and pore size distribution of nanocarbon structures allow them to increase the performance of electrochemical capacitance in terms of both the power delivery rate and the energy storage capacity. Some nanocarbon based materials used in the environmental and energy application are shown in **Table 1**.

4. Conclusion

This book chapter has focused on the application of nanocarbon-based materials/composites in the environmental and energy relevant area. Numerous exceptional properties of nanocarbon based such as, outstanding pore size distribution, large surface area, ease of porous texture modification, mechanical, thermal stability and chemical deformation make them appropriate for the different application. Overall functional group related to nanocarbon attached with specific materials/metals and increases the electrical, thermal and other desirable properties of the composite. Modified nanocarbon-based materials with enhanced electronics properties can be used for the different electronics devices in energy relevant area such as energy storage, conduction, radiation, etc., and environment relevant area such as, pollution parameter detecting devices. Higher surface area of nanocarbon based materials in comparison with conventional materials can be used in the pollution remediation application. Antibacterial nature of nanocarbon based materials can also be used in wastewater treatment for disinfection process and it can be used in preparation of biomedical relevant area to minimize bacterial contamination.

Author details

Prateek Khare, Ratnesh Kumar Patel and Ravi Shankar*
Department of Chemical Engineering, Madan Mohan Malaviya University of Technology, Gorakhpur, Uttar Pradesh, India

*Address all correspondence to: rsch@mmmut.ac.in

References

[1] Náray-Szabó G, Mika LT. Conservative evolution and industrial metabolism in green chemistry. Green Chemistry. 2018;**20**(10):2171-2191. DOI: 10.1039/C8GC00514A

[2] Kelly FJ, Fussell JC. Air pollution and public health: Emerging hazards and improved understanding of risk. Environmental Geochemistry and Health. 2015;**37**(4):631-649. DOI: 10.1007/s10653-015-9720-1

[3] Kopp RE, Mauzerall DL. Assessing the climatic benefits of black carbon mitigation. Proceedings of the National Academy of Sciences. 2010;**107**(26):11703-11708. DOI: 10.1073/pnas.0909605107

[4] Eskenazi B, Mocarelli P, Warner M, Needham L, Patterson DG, Samuels S, et al. Relationship of serum TCDD concentrations and age at exposure of female residents of Seveso, Italy. Environmental Health Perspectives. 2004;**112**(1):22-27. DOI: 10.1289/ehp.6573

[5] Schnitzer H, Ulgiati S. Less bad is not good enough: Approaching zero emissions techniques and systems. Journal of Cleaner Production. 2007;**15**(13-14):1185-1189. DOI: 10.1016/j.jclepro.2006.08.001

[6] Anastas PT, Kirchhoff MM. Origins, current status, and future challenges of green chemistry. Accounts of Chemical Research. 2002;**35**(9):686-694. DOI: 10.1021/ar010065m

[7] Zaytseva O, Neumann G. Carbon nanomaterials: Production, impact on plant development, agricultural and environmental applications. Chemical and Biological Technologies in Agriculture. 2016;**3**(1):17. DOI: 10.1186/s40538-016-0070-8

[8] Tripathi KM, Sonkar AK, Sonkar SK, Sarkar S. Pollutant soot of diesel engine exhaust transformed to carbon dots for multicoloured imaging of *E. coli* and sensing cholesterol. RSC Advances. 2014;**4**(57):30100-30107. DOI: 10.1039/C4RA03720K

[9] Uchida T, Ohashi O, Kawamoto H, Yoshimura H, Kobayashi K, Tanimura M, et al. Synthesis of single-wall carbon nanotubes from diesel soot. Japanese Journal of Applied Physics. 2006;**45**(10A):8027-8029. DOI: 10.1143/JJAP.45.8027

[10] Wang H, Ma H, Zheng W, An D, Na C. Multifunctional and recollectable carbon nanotube ponytails for water purification. ACS Applied Materials & Interfaces. 2014;**6**(12):9426-9434. DOI: 10.1021/am501810f

[11] Lee TH, Yao N, Hsu WK. Fullerene-like carbon particles in petrol soot. Carbon. 2002;**40**(12):2275-2279. DOI: 10.1016/S0008-6223(02)00198-7

[12] Singh A, Khare P, Verma S, Bhati A, Sonker AK, Tripathi KM, et al. Pollutant soot for pollutant dye degradation: Soluble graphene nanosheets for visible light induced photodegradation of methylene blue. ACS Sustainable Chemistry & Engineering. 2017;**5**(10):8860-8869. DOI: 10.1021/acssuschemeng.7b01645

[13] Khare P, Singh A, Verma S, Bhati A, Sonker AK, Tripathi KM, et al. Sunlight-induced selective photocatalytic degradation of methylene blue in bacterial culture by pollutant soot derived nontoxic graphene nanosheets. ACS Sustainable Chemistry & Engineering. 2018;**6**(1):579-589. DOI: 10.1021/acssuschemeng.7b02929

[14] Tripathi KM, Singh A, Bhati A, Sarkar S, Sonkar SK. Sustainable feasibility of the environmental pollutant soot to few-layer photoluminescent graphene nanosheets

for multifunctional applications. ACS Sustainable Chemistry & Engineering. 2016;**4**(12):6399-6408. DOI: 10.1021/acssuschemeng.6b01045

[15] Afreen S, Omar RA, Talreja N, Chauhan DA, Shfaq M. Carbon-based nano structured materials for energy and environmental remediation applications. In: Prasad R, Aranda E, editors. Approaches in Bioremediation, Nanotechnology in the Life Sciences. 1st ed. Switzerland: Springer; 2018. pp. 369-392. DOI: 10.1007/978-3-030-02369-0_17

[16] Commission E. Commission recommendations of 18 October 2011 on the definition of nanomaterial. Official Journal of the European Union. 2011;**54**:38-40

[17] De-Volder MF, Tawfick SH, Baughman RH, Hart AJ. Carbon nanotubes: Present and future commercial applications. Science. 2013;**339**(6119):535-539. DOI: 10.1126/science.1222453

[18] Khare R, Bose S. Carbon nanotube based composites—A review. Journal of Minerals and Materials Characterization and Engineering. 2005;**4**(1):31-46. DOI: 10.4236/jmmce.2005.41004

[19] Khare P, Bhati A, Anand SR, Gunture, Sonkar SK. Brightly fluorescent zinc-doped red-emitting carbon dots for the sunlight-induced photoreduction of Cr(VI) to Cr(III). ACS Omega. 2018;**3**(5):5187-5194. DOI: 10.1021/acsomega.8b00047

[20] Bhati A, Anand SR, Kumar G, Garg A, Khare P, Sonkar SK. Sunlight-induced photocatalytic degradation of pollutant dye by highly fluorescent red-emitting Mg-N-embedded carbon dots. ACS Sustainable Chemistry & Engineering. 2018;**6**(7):9246-9256. DOI: 10.1021/acsomega.8b00047

[21] Han M, Zhu S, Lu S, Song Y, Feng T, Tao S, et al. Recent progress on the photocatalysis of carbon dots: Classification, mechanism and applications. Nano Today. 2018;**19**:201-218. DOI: 10.1016/j.nantod.2018.02.008

[22] Kroto HW, Heath JR, O'Brien SC, Curl SF, Smalley. C60: Buckminsterfullerene. Nature. 1985;**318**(6042):162-163. DOI: 10.1038/318162a0

[23] Rysaeva LK. Mechanical properties of fullerite of various composition. Journal of Physics Conference Series. 2017;**938**(1):012071. DOI: 10.1088/1742-6596/938/1/012071

[24] Mousavi SZ, Nafisi S, Maibach HI. Fullerene nanoparticle in dermatological and cosmetic applications. Nanomedicine: Nanotechnology, Biology and Medicine. 2017;**13**(3):1071-1087. DOI: 10.1016/j.nano.2016.10.002

[25] Noh SH, Kwon C, Hwang J, Ohsaka T, Kim BJ, Kim TY, et al. Self-assembled nitrogen-doped fullerenes and their catalysis for fuel cell and rechargeable metal–air battery applications. Nanoscale. 2017;**9**(22):7373-7379. DOI: 10.1039/C7NR00930E

[26] Yi H, Huang D, Qin L, Zeng G, Lai C, Cheng M, et al. Selective prepared carbon nanomaterials for advanced photocatalytic application in environmental pollutant treatment and hydrogen production. Applied Catalysis B: Environmental. 2018;**239**:408-424. DOI: 10.1016/j.apcatb.2018.07.068

[27] Yadav BC, Srivastava R, Kumar A. Characterization of ZnO nanomaterial synthesized by different methods. International Journal of Nanotechnology and Applications. 2007;**1**:1-11

[28] Sharma AK, Khare P, Singh JK, Verma N. Preparation of novel carbon microfiber/carbon nanofiber-dispersed polyvinyl alcohol-based nanocomposite

material for lithium-ion electrolyte battery separator. Materials Science and Engineering: C. 2013;**33**(3):1702-1709. DOI: 10.1016/j.msec.2012.12.083

[29] Khare P, Ramkumar J, Verma N. Control of bacterial growth in water using novel laser-ablated metal–carbon–polymer nanocomposite-based microchannels. Chemical Engineering Journal. 2015;**276**:65-74. DOI: 10.1016/j.cej.2015.04.060

[30] Khare P, Yadav A, Ramkumar J, Verma N. Microchannel-embedded metal–carbon–polymer nanocomposite as a novel support for chitosan for efficient removal of hexavalent chromium from water under dynamic conditions. Chemical Engineering Journal. 2016;**293**:44-54. DOI: 10.1016/j.cej.2016.02.049

[31] Zhang R, Zhang Y, Zhang Q, Xie H, Qian W, Wei F. Growth of half-meterlong carbon nanotubes based on Schulz-Flory distribution. ACS Nano. 2013;**7**(7):6156-6161. DOI: 10.1021/nn401995z

[32] Tang B, Chen H, Peng H, Wang Z, Huang W. Graphene modified TiO_2 composite photocatalysts: Mechanism, progress and perspective. Nanomaterials. 2018;**8**(2):105. DOI: 10.3390/nano8020105

[33] Xu J, Cao Z, Zhang Y, Yuan Z, Lou Z, Xu X, et al. A review of functionalized carbon nanotubes and graphene for heavy metal adsorption from water: Preparation, application, and mechanism. Chemosphere. 2018;**195**:351-364. DOI: 10.1016/j.chemosphere.2017.12.061

[34] Xu X, Ray R, Gu Y, Ploehn HJ, Gearheart L, Raker K, et al. Electrophoretic analysis and purification of fluorescent single-walled carbon nanotube fragments. Journal of the American Chemical Society. 2004;**126**(40):12736-12737. DOI: 10.1021/ja040082h

[35] Sun YP, Zhou B, Lin Y, Wang W, Fernando KAS, Pathak P, et al. Quantum-sized carbon dots for bright and colorful photoluminescence. Journal of the American Chemical Society. 2006;**128**(24):7756-7757. DOI: 10.1021/ja062677d

[36] Wu ZL, Liu ZX, Yuan YH. Carbon dots: Materials, synthesis, properties and approaches to long-wavelength and multicolor emission. Journal of Materials Chemistry B. 2017;**5**(21):3794-3809. DOI: 10.1039/C7TB00363C

[37] Gao X, Du C, Zhuang Z, Chen W. Carbon quantum dot-based nanoprobes for metal ion detection. Journal of Materials Chemistry C. 2016;**4**(29):6927-6945. DOI: 10.1039/C6TC02055K

[38] Liu W, Li C, Ren Y, Sun X, Pan W, Li Y, et al. Carbon dots: Surface engineering and applications. Journal of Materials Chemistry B. 2016;**4**(35):5772-5788. DOI: 10.1039/C6TB00976J

[39] Xu Q, Kuang T, Liu Y, Cai L, Peng X, Sreeprasad TS, et al. Heteroatom-doped carbon dots: Synthesis, characterization, properties, photoluminescence mechanism and biological applications. Journal of Materials Chemistry B. 2016;**4**(45):7204-7219. DOI: 10.1039/C6TB02131J

[40] Sharma V, Tiwari P, Mobin SM. Sustainable carbon-dots: recent advances in green carbon dots for sensing and bioimaging. Journal of Materials Chemistry B. 2017;**5**(45):8904-8924. DOI: 10.1039/C7TB02484C

[41] MacKenzie KJ, See CH, Dunens OM, Harris AT. Do single-walled carbonnanotubes occur naturally? Nature Nanotechnology. 2008;**3**:310. DOI: 10.1038/nnano.2008.139

[42] Velasco-Santos C, Martinez-Hernandez AL, Consultchi A,

Rodriguez R, Castano VM. Naturally produced carbon nanotubes. Chemical Physics Letters. 2003;**373**:272-276. DOI: 10.1016/S0009-2614(03)00615-8

[43] Su DS, Chen X. Natural lavas as catalysts for efficient production of carbon nanotubes and nanofibers. Angewandte Chemie, International Edition. 2007;**46**(11):1823-1824. DOI: 10.1002/anie.200604207

[44] Mracek J, Fagan RD, Stengelin RM, Hesjedal T. Are carbon nanotubes a naturally occurring material? Hints from methane CVD using lava as acatalyst. Current Nanoscience. 2011;**7**(3):294-296. DOI: 10.2174/157341311795542543

[45] Buseck PR, Tsipursky SJ, Hettich R. Fullerenes from the geological environment. Science. 1992;**257**(5067):215-217. DOI: 10.1126/science.257.5067.215

[46] Becker L, Bada JL, Winans RE, Bunch TE. Fullerenes in Allende meteorite. Nature. 1994;**372**:507. DOI: 10.1038/372507a0

[47] Tran TH, Nguyen HL, Hwang DS, Lee JY, Cha HG, Koo JM, et al. Five different chitin nanomaterials from identical source with different advantageous functions and performances. Carbohydrate Polymers. 2019;**205**:392-400. DOI: 10.1016/j.carbpol.2018.10.089

[48] Okan M, Aydin HM, Barsbay M. Current approaches to waste polymer utilization and minimization: A review. Journal of Chemical Technology and Biotechnology. 2019;**94**(1):8-21. DOI: 10.1002/jctb.5778

[49] Hirsch A, Vostrowsky O. Functionalization of carbon nanotubes. Topics in Current Chemistry. 2005;**245**:193-237. DOI: 10.1007/b98169

[50] Hernández-Fernández P, Montiel M, Ocón P, Fuente JL, García-Rodríguez S, Rojas S, et al. Functionalization of multi-walled carbon nanotubes and application as supports for electrocatalysts in proton-exchange membrane fuel cell. Applied Catalysis B: Environmental. 2010;**99**(1-2):343-352. DOI: 10.1016/j.apcatb.2010.07.005

[51] Li X, Qin Y, Picraux ST, Guo ZX. Noncovalent assembly of carbon nanotube-inorganic hybrids. Journal of Materials Chemistry. 2011;**21**:7527-7547. DOI: 10.1039/c1jm10516g

[52] Jiang L, Gao L, Sun J. Production of aqueous colloidal dispersions of carbon nanotubes. Journal of Colloid and Interface Science. 2003;**260**(1):89-94. DOI: 10.1016/S0021-9797(02)00176-5

[53] Ahmad K, Pan W. Microstructure-toughening relation in alumina based multiwall carbon nanotube ceramic composites. Journal of the European Ceramic Society. 2015;**35**(2):663-671. DOI: 10.1016/j.jeurceramsoc.2014.08.044

[54] Nasibulin AG, Shandakov SD, Nasibulina LI, Cwirzen A, Mudimela PR, Habermehl-Cwirzen K, et al. A novel cement-based hybrid material. New Journal of Physics. 2009;**11**:023013. DOI: 10.1088/1367-2630/11/2/023013

[55] Ma PC, Zhang Y. Perspectives of carbon nanotubes/polymer nanocomposites for wind blade materials. Renewable and Sustainable Energy Reviews. 2014;**30**:651-660. DOI: 10.1016/j.rser.2013.11.008

[56] Loh KJ, Ryu D. 11-multifunctional materials and nanotechnology for assessing and monitoring civil infrastructures. In: Wang M, Lynch J, Sohn H, editors. Sensor technologies for civil infrastructures: Woodhead Publishing Series in Electronic and Optical Materials. Vol. 1. Sawston: Woodhead Publishing; 2014. pp. 295-326. DOI: 10.1533/9780857099136.295

[57] Ng K, Lam W, Pichiah S. A review on potential applications of carbon nanotubes in marine current turbines. Renewable and Sustainable Energy Reviews. 2013;**28**:331-339. DOI: 10.1016/j.rser.2013.08.018

[58] Tan D, Zhang Q. Research of carbon nanotubes/polymer composites for sports equipment. In: Zhang T, editor. Future Computer, Communication, Control and Automation. Advances in Intelligent and Soft Computing. Vol. 119. Berlin, Heidelberg: Springer; 2012. pp. 137-146. DOI: 10.1007/978-3-642-25538-0_20

[59] Peddini SK, Bosnyak CP, Henderson NM, Ellison CJ, Paul DR. Nanocomposites from styrene-butadiene rubber (SBR) and multiwall carbon nanotubes (MWCNT) part 1: Morphology and rheology. Polymer. 2014;**55**(1):258-270. DOI: 10.1016/j.polymer.2013.11.003

[60] Gohardani O, Elola MC, Elizetxea C. Potential and prospective implementation of carbon nanotubes on next generation aircraft and space vehicles: A review of current and expected applications in aerospace sciences. Progress in Aerospace Sciences. 2014;**70**:42-68. DOI:10.1016/j.paerosci.2014.05.002

[61] Uritu CM, Varganici CD, Ursu L, Coroaba A, Nicolescu A, Dascalu AI, et al. Hybrid fullerene conjugates as vectors for DNA cell-delivery. Journal of Materials Chemistry B. 2015;**3**: 2433-2446. DOI: 10.1039/C4TB02040E

[62] Choi W, Lahiri I, Seelaboyina R, Kang YS. Synthesis of graphene and its applications: A review. Critical Reviews in Solid State and Materials Sciences. 2010;**35**(1):52-71. DOI: 10.1080/10408430903505036

[63] Lei T, Pochorovski I, Bao Z. Separation of semiconducting carbon nanotubes for flexible and stretchable electronics using polymer removable method. Accounts of Chemical Research. 2017;**50**(4):1096-1104. DOI: 10.1021/acs.accounts.7b00062

[64] Watanabe R, Matsuzaki R, Endo H, Koyanagi J. Stretchable and insulating characteristics of chemically bonded graphene and carbon nanotube composite materials. Journal of Materials Science. 2018;**53**(2):1148-1156. DOI: 10.1007/s10853-017-1563-y

[65] Shankar R, Singh L, Mondal P, Chand S. Removal of COD, TOC, and color from pulp and paper industry wastewater through electrocoagulation. Desalination and Water Treatment. 2014;**52**(40-42):7711-7722. DOI: 10.1080/19443994.2013.831782

[66] Shankar R, Varma AK, Mondal P, Chand S. Simultaneous treatment and energy production from PIW using electro coagulation & microbial fuel cell. Journal of Environmental Chemical Engineering. 2016;**4**(4A):4612-4618. DOI: 10.1016/j.jece.2016.10.021

[67] Vukovic GD, Marinkovic AD, Colic M, Ristic MD, Aleksic R, Peric-Grujic AA, et al. Removal of cadmium from aqueous solutions by oxidized and ethylenediamine-functionalized multi-walled carbon nanotubes. Chemical Engineering Journal. 2010;**157**(1):238-248. DOI: 10.1016/j.cej.2009.11.026

[68] Vukovic GD, Marinkovic AD, Skapin SD, Ristic MD, Aleksic R, Peric-Grujic AA, et al. Removal of lead from water by amino modified multiwalled carbon nanotubes. Chemical Engineering Journal. 2011;**173**(3): 855-865. DOI: 10.1016/j.cej.2011.08.036

[69] Yadav A, Verma N. Efficient hydrogen production using Ni-graphene oxide-dispersed laser-engraved 3D carbon micropillars as electrodes for microbial

electrolytic cell. Renewable Energy. 2019;**38**:628-638. DOI: 10.1016/j.renene.2019.01.100

[70] Khare P, Ramkumar J, Verma N. Carbon nanofiber-skinned three dimensional Ni/carbon micropillars: High performance electrodes of a microbial fuel cell. Electrochimica Acta. 2016;**219**:88-98. DOI: 10.1016/j.electacta.2016.09.140

[71] Smith SC, Rodrigues DF. Carbon-based nanomaterials for removal of chemical and biological contaminants from water: A review of mechanisms and applications. Carbon. 2015;**91**:122-143. DOI: 10.1016/j.carbon.2015.04.043

[72] Yan H, Gong A, He H, Zhou J, Wei Y, Lv L. Adsorption of microcystins by carbon nanotubes. Chemosphere. 2006;**62**(1):142-114. DOI: 10.1016/j.chemosphere.2005.03.075

[73] Li Y, Wang S, Wei J, Zhang X, Xu C, Luan Z, et al. Lead adsorption on carbon nanotubes. Chemical Physics Letters. 2002;**357**(3-4):263-266. DOI: 10.1016/S0009-2614(02)00502-X

[74] Dichiara AB, Webber MR, Gorman WR, Rogers RE. Removal of copper ions from aqueous solutions via adsorption on carbon nanocomposites. ACS Applied Materials & Interfaces. 2015;**7**(28):15674-15680. DOI: 10.1021/acsami.5b04974

[75] Zhang L, Song X, Liu X, Yang L, Pan F, Lv J. Studies on the removal of tetracycline by multi-walled carbon nanotubes. Chemical Engineering Journal. 2011;**178**:26-33. DOI: 10.1016/j.cej.2011.09.127

[76] Deng J, Shao Y, Gao N, Deng Y, Tan C, Zhou S, et al. Multiwalled carbon nanotubes as adsorbents for removal of herbicide diuron from aqueous solution. Chemical Engineering Journal. 2012;**193-194**:339-347. DOI: 10.1016/j.cej.2012.04.051

[77] Zheng X, Su Y, Chen Y, Wei Y, Li M, Huang H. The effects of carbon nanotubes on nitrogen and phosphorus removal from real wastewater in the activated sludge system. RSC Advances. 2014;**4**:45953-45959. DOI: 10.1039/C4RA04128C

[78] Jung JH, Hwang GB, Lee JE, Bae GN. Preparation of airborne Ag/CNT hybrid nanoparticles using an aerosol process and their application to antimicrobial air filtration. Langmuir. 2011;**27**(16):10256-10264. DOI: 10.1021/la201851r

[79] Al-Hakami SM, Khalil AB, Laoui T, Atieh MA. Fast disinfection of Escherichia colibacteria using carbon nanotubes interaction with microwave radiation. Bioinorganic Chemistry and Applications. 2013;**2013**:458943. DOI: 10.1155/2013/458943

[80] Arabi S, Lalehloo RS, Olyai M, Ali G, Sadegh H. Removal of congo red azo dye from aqueous solution by ZnO nanoparticles loaded on multiwall carbon nanotubes. Physica E: Low-Dimensional Systems and Nanostructures. 2019;**106**:150-155. DOI: 10.1016/j.physe.2018.10.030

[81] Zulkifili ANB, Kento T, Daiki M, Fujiki A. The basic research on the dye-sensitized solar cells (DSSC). Journal of Clean Energy Technologies. 2015;**3**(5):382-387. DOI: 10.7763/JOCET.2015.V3.228

[82] Yun S, Hagfeldt A, Ma T. Pt-free counter electrode for dye-sensitized solar cells with high efficiency. Advanced Materials. 2014;**26**(36):6210-6237. DOI: 10.1002/adma.201402056

[83] Hwang S, Batmunkh M, Nine MJ, Chung H, Jeong H. Dye-sensitized solar cell counter electrodes based on carbon nanotubes. ChemPhysChem. 2015;**16**(1):53-65. DOI: 10.1002/cphc.201402570

[84] Zheng X, Zhang L. Photonic nanostructures for solar energy conversion. Energy & Environmental Science. 2016;**9**:2511-2532. DOI: 10.1039/C6EE01182A

[85] Wu Z, Li L, Yan JM, Zhang XB. Materials design and system construction for conventional and new concept super capacitors. Advanced Science. 2017;**4**(6):1600382. DOI: 10.1002/advs.201600382

[86] Lukatskaya MR, Dunn B, Gogotsi Y. Multidimensional materials and device architectures for future hybrid energy storage. Nature Communications. 2016;**7**:12647. DOI: 10.1038/ncomms12647

[87] Zuo W, Li R, Zhou C, Li Y, Xia J, Liu J. Battery-supercapacitor hybrid devices: recent progress and future prospects. Advanced Science. 2017;**4**(7):1600539. DOI: 10.1002/advs.201600539

[88] Zheng JS, Zhang L, Shellikeri A, Cao W, Wu Q, Zheng JP. A hybrid electrochemical device based on a synergetic inner combination of Li ion battery and Li ion capacitor for energy storage. Scientific Reports. 2017;**7**:41910. DOI: 10.1038/srep41910

[89] Wang J, Kaskel S. KOH activation of carbon-based materials for energy storage. Journal of Materials Chemistry. 2012;**22**:23710-23725. DOI: 10.1039/C2JM34066F

[90] Ma J, Yang M, Yu F, Chen J. Easy solid-phase synthesis of pH-insensitive heterogeneous CNTs/FeS Fenton-like catalyst for the removal of antibiotics from aqueous solution. Journal of Colloid and Interface Science. 2015;**444**:24-32. DOI: 10.1016/j.jcis.2014.12.027

[91] Hummers WS, Offeman RE. Preparation of graphitic oxide. Journal of the American Chemical Society. 1958;**80**(6):1339-1339. DOI: 10.1021/ja01539a017

[92] Thakur S, Das G, Raul PK, Karak N. Green one-step approach to prepare sulfur/reduced graphene oxide nanohybrid for effective mercury ions removal. The Journal of Physical Chemistry C. 2013;**117**(15):7636-7642. DOI: 10.1021/jp400221k

[93] Xu Y, Bai H, Lu G, Li C, Shi G. Flexible graphene films via the filtration of water-soluble noncovalent functionalized graphene sheets. Journal of The American Chemical Society. 2008;**130**(18):5856-5857. DOI: 10.1021/ja800745y

[94] Qi GQ, Yang J, Bao RY, Liu ZY, Yang W, Xie BH, et al. Enhanced comprehensive performance of polyethylene glycol based phase change material with hybrid graphene nanomaterials for thermal energy storage. Carbon. 2015;**88**:196-205. DOI: 10.1016/j.carbon.2015.03.009

[95] Yan Y, Zheng W, Su L, Mao L. Carbon-nanotube-based glucose/O_2 biofuel cells. Advanced Materials. 2006;**18**(19):2639-2643. DOI: 10.1002/adma.200600028

[96] Yu S, Jeong SG, Chung O, Kim S. Bio-based PCM/carbon nanomaterials composites with enhanced thermal conductivity. Solar Energy Materials & Solar Cells. 2014;**120**(B):549-554. DOI: 10.1016/j.solmat.2013.09.037

[97] Cui G, Zhi L, Thomas A, Lieberwirth I, Kolb U, Müllen K. A novel approach towards carbon–Ru electrodes with mesoporosity for supercapacitors. ChemPhysChem. 2007;**8**(7):1013-1015. DOI: 10.1002/cphc.200600789

Chapter 2

Synthesis and Characterization of Nanocomplexes by Green Chemistry and Their Applications in Different Fields

Rania H. Taha

Abstract

In this study, a convenient synthesis of a series of La(III) complexes (in nanoscales) with different organic ligands was prepared by green chemical method. The physicochemical properties of the nanocomplexes were obtained. The SEM and TEM techniques were used to confirm the nanosize and morphology. The cytotoxicity of the newly synthesized compounds was demonstrated and tested for their anticancer activity against common pathogenic organisms and against human epithelial colorectal adenocarcinoma cells (Caco2), Vero, and MCF-7 cells. By examining its pharmacological activity, we were able to identify new potent and selective anticancer agents. Also, the structures succeeded to affect and stop the growth of *Klebsiella pneumonia* strain, and the results revealed promising activity. In addition, an electrospinning technique was used for the fabrication of novel bio-medicated nanofibers which are applied for wound dressing as local chemotherapy for skin injuries.

Keywords: nanocomplexes, anticancer activity, Caco2, vero, MCF-7 cells, flow cytometry, electrospinning, wound dressing

1. Introduction

The Environmental Protection Agency defined the green chemistry or sustainable chemistry as "the production of chemical products and processes which have the ability to reduce or eliminate the use or generation of hazardous substances." In recent years there is an expectation that chemists and chemical engineers should design greener and more sustainable chemical processes, and it was expected that this trend will continue to grow over the next few decades [1].

Studies on metal-based drugs offer promise in the fight against cancer, which is the main aim of the inorganic chemistry [2, 3].

The metal-based anticancer drugs are driven by the necessity to fill the gap in tumor chemotherapy in order to minimize the undesirable side effects and enlarge the spectrum of activity for more tumor types and metastatic cancers. The Schiff base structures play a key role in studying the mechanism of the transformation and racemization reaction in biological systems due to the similarity of their structures to the natural biological system [4–11].

Among various metal complexes, La(III) complexes have been intensively investigated due to their more physiological activities and lower toxicities after coordination with ligand, so, in recent years people have paid great interest to synthesis, DNA interaction, and anticancer activity of La(III) complexes [12–16].

Dressings play multimodal roles in wound healing process, such as preventing infections, absorbing exudates, and maintaining moisture, and thus are indispensable [17]. Electrospinning is the method to fabricate ultrafine fibers from polymer solution(s) or melt(s) with diameters ranging from nanometers to micrometers [18].

The obtained results revealed the successful implementation of the nanoparticle-nanofiber as a local wound dressing especially for skin injuries.

2. Experimental

2.1 Materials and reagents

In this study all chemicals used were of highest purity. Organic solvents such as ethyl alcohol and dimethylformamide (DMF) were spectroscopically pure from the British Drug House (BDH). The other materials were purchased from Sinopharm Chemical Reagent Co., Ltd. and used without further purification. In addition to the 5-bromosalicylaldehyde (Sigma-Aldrich), $La(NO_3)_3{\cdot}6H_2O$ (Merck) were also used.

2.2 Instrumentation

In open capillaries melting points (°C, uncorrected) were determined on a Gallenkamp melting point apparatus (Sanyo Gallenkamp, Southborough, UK). In the Microanalytical Center, Cairo University (C, H, N, and S) were carried out. ^{1}H NMR spectra (DMSO-d_6) were measured at Bruker FT-400MHZ spectrophotometer. The IR spectra were recorded on a Perkin Elmer 437 IR spectrophotometer (400–4000 cm^{-1}) (KBr technique). Thermogravimetric analysis was performed using a Shimadzu TGA-50H with a flow rate of 20 ml min^{-1}. The UV-vis spectra were recorded on a Perkin Elmer Lambda 3B UV-vis spectrophotometer. X-ray diffraction were recorded at room temperature (~25°C) on Empyrean X-ray diffractometer equipped. The patterns were run with Cu target (CuKα radiation), and the tube operated at 45 kV and 30 mA. The size and morphology of the nanocomplexes were characterized with a scanning electron microscope (SEM) (Philips XL 30) with gold coating and TEM.

2.3 Cytotoxicity assay

Different concentrations of the nanometal complexes were tested for their cytotoxicity against Vero, Caco2, and MCF-7 cell lines using (MTT) Thiazolyl Blue Tetrazolium Bromide method according to [19, 20]. In brief, Vero, Caco2, and MCF-7 cells (10×10^3) were cultured in a 96-well plate overnight at 37°C, 5% CO_2, and 88% humidity. The total volume of used DEMEM supplemented medium and the synthesized compound supernatant was 200 μL with final concentrations of 10, 20, 30, 40, 60, 80, and 100 mg/L. The plate was incubated at 37°C and 5% CO_2 for 3 days. After incubation, debris and dead cells were removed by washing three times with fresh medium. Twenty microliters of MTT solution (5 mg/mL of MTT in PBS buffer) was added to each well and shook for 5 min at 150 rpm to thoroughly mix the MTT into the media. The cells were incubated at 37°C and 5% CO_2 for 3–5 h to metabolize MTT by viable cells. Two hundred microliter dimethyl sulfoxide (DMSO) was added to each well and shook again for 5 min at 150 rpm, and then the viability of the cells was calculated by measuring the optical density at 630 nm subtracted from optical density at 570 nm [21]. The percentage of viability cells was calculated by comparison with control cells (without adding the synthesized compounds to the cells) using the equation (A) test/(A) control $\times$ 100.

2.4 Flow cytometry

To discriminate living cells from dead cells or for cell cycle analysis, propidium iodide (PI) can be used according to Léonce et al. [22]. This analysis is based on the stoichiometric binding of PI to intracellular DNA. At the end, cells were washed with PBS and collected by trypsinization. Cells were then resuspended in warm PBS and fixed with 4 ml ice-cold ethanol. Finally, in darkness, cells were stained with 0.5 ml of warm PI solution; 7 ml of PI solution consists of 0.35 ml of PI stock solution (1 mg/ml), 0.7 ml RNase A stock solution (1 mg/ml), and 6 ml of PBS and incubated in darkness for 30 minutes. The samples were kept on ice until flow cytometric analysis.

2.5 Preparation of (chitosan-La(III) nanocomplex) composite nanofibers

Chitosan (CS) was dissolved in 50% acetic acids at room temperature under moderate stirring for 12 h to form a homogeneous solution, and then the solution of La complex nanoparticles was quietly adding. The total solution was fixed to 3 wt.%. The solution was filled in a 20 ml NORM-JECT Luer Lock tip plastic syringe having an 18-gauge stainless steel needle with 90° blunt end. The electrospinning setup included a high voltage power supply, purchased from the NanoNC, Inc. (S. Korea), and a nanofiber collector of aluminum foil that covered a laboratory-produced roller with the diameter of 10 cm. The collector was placed at 10 cm tip-to-collector distance (TCD) [39].

2.6 Synthesis

2.6.1 Synthesis of complex nanoparticles by sonochemistry method (green chemistry)

Ten milliliter of 0.1 M solution of $La(NO_3)_3.6H_2O$ in ethanol was positioned in a high-density ultrasonic probe, operating at 24 kHz with a maximum power [23] output of 400 W. Ten milliliter of 0.1 M solution of N'-((E)-5-bromo-2-hydroxybenzylidene)-6-(((E)-5-bromo-2-hydroxybenzylidene)amino)-4-oxo-2-thioxo-1,2,3,4-tetrahydrothieno[3,2-d]pyrimidine-7-carbohydrazide H_2L^1, N-(2-mercaptophenyl) benzamide H_2L^2, and 2-((2-mercapto- phenyl) imino)-1,2-diphenylethan-1-ol H_2L^3 was added dropwise. The obtained precipitates were allowed to evaporate at room temperature to obtain the complex nanoparticle in powder form.

3. Results and discussion

3.1 Elemental analyses for nanometal complexes

In the present investigation, the analytical data of the prepared metal complexes suggest the structures as given in **Figure 6**. The physicochemical results of the newly synthesized compounds were presented in **Table 1**. The obtained results were in good agreement with those calculated for the suggested formulae. The melting points were sharp, indicating the purity of the prepared nanometal complexes. The proposed formula of the nanometal complexes is according to the following general equations:

$$La(NO_3)_3.6H_2O + H_2L^{1\text{-}3} \longrightarrow \left[(La)(L^{1\text{-}3})nNO_3.xH_2O\right]yH_2O.zNO_3$$

Where

at H_2L^1 : n = 1, x = 1, y = 0, z = 0

H_2L^2 : n = 2, x = 2, y = 1, z = 1

H_2L^3 : n = 2, x = 2, y = 0, z = 0

3.2 ^{1}H NMR spectra

The ^{1}HNMR spectra (**Figure 1**, **Table 2**) of the La (III) nanocomplex **1** showed a singlet peak at 8.00 ppm assigned to azomethine proton, in addition to the multiplet signals due to aromatic protons. It was found that the spectra of complex **2** display singlet peaks due to SH and NH groups at δ3.5 and 5.4 ppm indicating the involvement of these groups in coordination with the metal atoms without deprotonation.

In addition to the ^{1}H NMR spectra of complex **3**, showed singlet peaks due to the O–H proton at 10.4 ppm and the S–H proton peak at 3.47 ppm. This analysis showed that the O–H and S–H groups share in the complexation a loss of their protons. Moreover, the spectra of the complexes showed multiplet signals at 6.99–7.48 ppm attributed to the aromatic protons for all complexes [24, 25].

Compd. code	(Empirical formula)	M.P. (°C)	Color (yield %)	(%) found (calcd.)				
				C	H	N	S	Λm*
1	$[(La)(L^1)(NO_3)(H_2O)]$	<350	Faint orange (80.54)	29.90 (30.01)	1.32 (1.56)	9.92 (10.00)	7.58 (7.62)	20.51
2	$[La(H_2L^2).(NO_3)_2(H_2O)_2]NO_3.H_2O$	110	Dark gray (85.49)	2.94 (3.06)	27.75 (27.87)	5.40 (5.71)	9.80 (10.00)	57.88
3	$[La(H_2L^3)(NO_3)_3.H_2O]$	165	Page (79.80)	2.83 (2.90)	36.54 (36.31)	4.55 (4.83)	8.33 (8.47)	3.39

*$ohm^{-1}cm^2mol^{-1}$

Table 1.
Elemental analysis and some physical measurements of La(III) nanocomplexes (1-3).

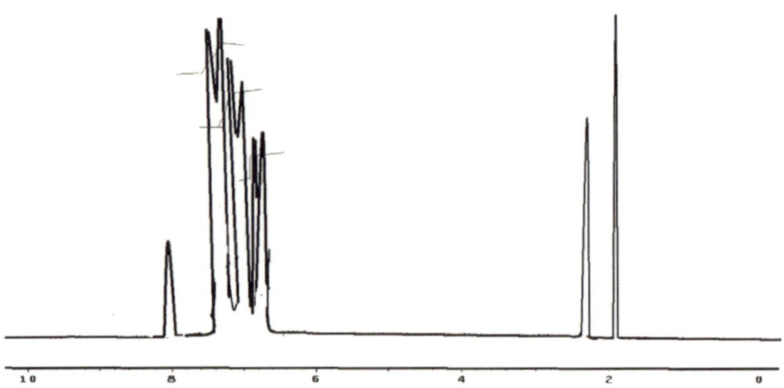

Figure 1.
^{1}H NMR data for La(III) diamagnetic complex (1).

Compd. no.	δ_{ph}	$\delta_{CH=N}$	δ_{SH}	δ_{NH}	δ_{OH}
1	6.39–8.46	8.00	–	–	–
2	6.99–7.48	–	3.5	5.4	–
3	7.00–8.00	–	3.47	–	10.4

Table 2.
^{1}H NMR data for La(III) nanocomplexes.

3.3 IR spectra and mode of bonding

In order to ascertain the mode of bonding, the IR spectrum of the free ligands was compared with those of their La complexes. By careful comparison of the spectra of the complexes with those of the free ligands, it was found that the band at 3430 cm^{-1} is found to be disappeared by complexation suggesting the involvement of OH groups of the phenolic groups in complex formation with their deprotonation. The bands due to NH and SH are shifted to higher or lower frequencies indicating the participation of –N and –S donors in the coordination, while the bands due to C=O are unchanged. The strong bands at 1622 and 1298 cm^{-1} due to $\upsilon_{C=N}$ and υ_{C-O} (azomethine and phenolic groups) are shifted to lower wave number 1609–1565 and 1280–1242 cm^{-1}, respectively, in metal complexes indicating the participation of nitrogen of azomethine and oxygen of phenolic group in complexation. The complexes showing also three bands at the range 1480–1400, 1380–1350, and 1170–1020 cm^{-1} indicate the unidentate coordination mode of the nitrate group. The curves also, showed that shift in the peaks of O–H, C–O, C–S groups (complex **3**) due to their sharing in the complexation and that recorded by the M–O, and M–S peaks. All these results are in good agreement with the conductance data. Another evidence for the coordination is the observance of new bands in the far IR region at 570–505, 520–440, and 432–524 cm^{-1} which may be due to υ_{M-O} υ_{M-N} and υ_{M-S}, respectively, that are not observed in the spectrum of the free ligands [26, 27].

3.4 Molar conductivity measurements

The conductivity Λ_m value of the La nanocomplexes **1–3** can be calculated by using the relation Λ_m = K/C. The complexes were (10^{-3} M) dissolved in DMF, and the molar conductivities of their solutions at 25±2°C were measured (**Table 1**). It is concluded from the results that complexes 1 and 3 are found to have molar conductance values of 20.51 and 3.39 ohm^{-1} mol^{-1} cm^2, respectively, indicating that these complexes are nonelectrolytic and monomeric in nature. Also the values indicate the bonding of the nitrate ions to metal cations [28, 29]. The molar conductivity value of complex **2** is seen to be 57.88 ohm^{-1} mol^{-1} cm^2 indicating the ionic nature of this complex, and the nitrate ion is outside the coordination sphere.

3.5 Electronic spectral data

The assignments of the significant electronic spectral absorption band of the nanometal complexes are given in Table 3. Two absorption peaks were observed in the spectra range of 210–227 and 279–337 nm due to п-п* and n-п* transitions, respectively, due to benzene and the azomethine (CH=N) function [26, 29]. Moreover, the spectra of complexes indicate no significant absorption in the visible region due to the absence of *f–f* transition, since *f–f* transitions are Laporte-forbidden and very weak in nature.

Compd. no.	μ_{eff} (B.M.)	Absorption bands (nm)	
		π-π*	n-π*
1	Diamagnetic	269	337
2	Diamagnetic	210	280
3	Diamagnetic	227	279

Table 3.
Magnetic moment and electronic spectral data of La(III) nanocomplexes.

3.6 Magnetic susceptibility

The following equation was used to calculate the magnetic susceptibility of the La nanocomplexes:

$$X_g = \frac{C(bal)(1)(R - R^0)}{10^{-9}\text{m}} \quad (1)$$

where C, calibration constant of standard; l, length of sample in filled tube; R^0, reading of empty sample tube; R, reading with sample; χ_g gram magnetic susceptibility.

The gram magnetic susceptibility of the complexes was found in the range -2.2098×10^{-6} –(-2.3013 $\times 10^{-6}$) $cm^3\ mol^{-1}$, respectively. Negative values of χ_g show that these complexes are diamagnetic in nature or there is no significant magnetic moment (**Table 3**).

3.7 Thermal analysis

Thermogravimetric analyses (TGA) of the La nanocomplexes were used to determine the thermal stability of the complexes, decide whether the water molecules are lattice or coordinated, and finally suggest a scheme for thermal decomposition of these chelates. In the present investigation, heating rates were suitably controlled at 10°C min^{-1} under nitrogen atmosphere, and the weight loss was measured from the ambient temperature up to ~ 600°C. The weight loss for each chelate was calculated within the corresponding temperature ranges.

Also, TGA curve of the La(III) complex of nanoscale **$[(La)(L^1)(NO_3)(H_2O)]$ (1)** showed three thermal stages. The first stage represents the loss of H_2O molecule of coordination, HNO_3 and ½O_2 with a found weight loss of 11.72% (cacld. 11.54%) within the temperature range of 130–180°C; the subsequent second step (180–300°C) corresponds to the loss of $C_5H_3BrN_2O$ with an estimated mass loss of 22.60% (calcd. 22.27%). The final stage at 300–520°C corresponds to loss of rest of the organic part ($C_{12}H_7BrN_3O_2S_2$) with an estimated weight loss 44.25% (calcd. 43.94%) leaving 4 C and La metal as a metallic residue. The overall weight loss amounts is 78.57% (cacld. 77.75%):

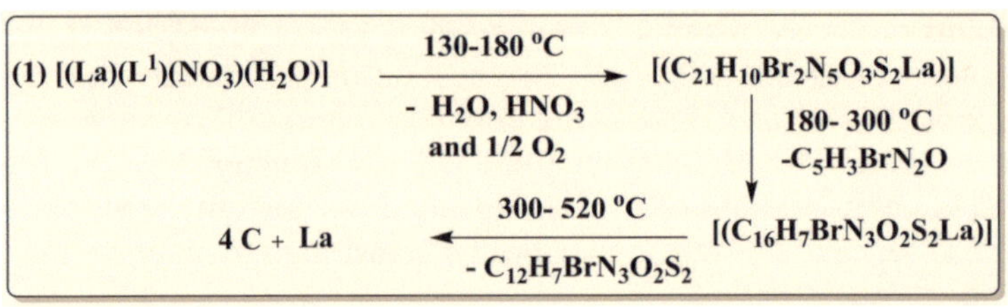

3.8 Powder X-ray diffraction studies

The diffraction of X-rays from the planes of a crystal (diffraction analysis) is considered as one of the important methods of the analytical work. By using X-ray diffraction which depends on the crystal properties of solids, we can identify the crystal structure of various solid compounds and identify the actual compounds from its structure, and we can also determine the arrangement of molecules in crystal [30]. X-ray diffractometry is an important technique as it is a nondestructive, non-contrast, fast, and sensitive one. Obtained XRD data of H_2L^1 ligand and its La(III) complex **1** of nanoparticle (**Figure 2**) show that XRD pattern of the ligand is different in comparison with the XRD pattern of the nanocomplex. The obtained data indicate that the complex is more crystalline in nature than that of the ligand which has an amorphous phase. The average particle size of the complex (1) can be estimated from Debye-Scherrer equation (Eq. (2)):

$$L = K\lambda/\beta \cos\theta \tag{2}$$

where β the breadth of the observed diffraction line at its half intensity maximum; K, the shape factor, 0.89; λ, the wavelength of the X-ray source used in XRD; θ, the corresponding incidence angle.

The average particle size of the nanoparticles is 15 nm. The sharp diffraction peaks of the samples indicate that the well-crystallized lanthanum complex nanocrystals can be prepared by our procedure [23].

3.9 EDX spectra

The amount of elements present in the percentage level of the metal complexes was identified by EDX data [31]. EDX spectra are used to calculate the percentage level of the elements present in the metal complexes like C, O, N, S, and La that are present in the La(III) nanocomplex **2**, shown in **Figure 3**. The revealed data are in good agreement with that of the elemental analysis.

3.10 SEM analysis

SEM technique was used to study the morphology and size of nanoparticles (**Figure 4**). It seems that the particles are semispherical with some agglomerations

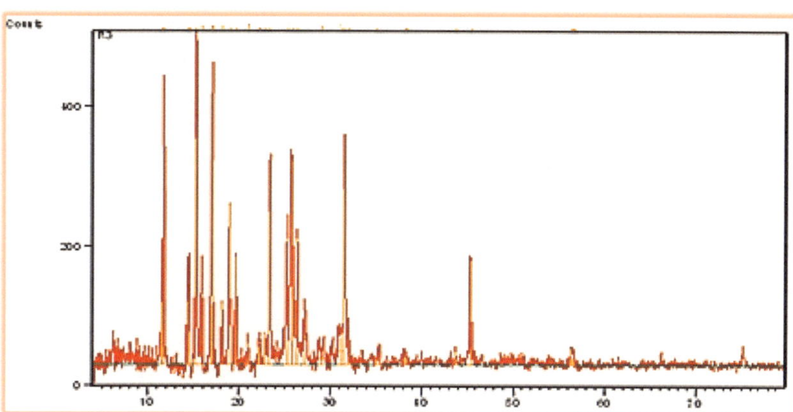

Figure 2.
X-ray powder diffractogram of complex (1) nanoparticle.

of the particles. The SEM images further revealed the stabilization of La(III) **1** nanoparticles due to interaction with the Schiff base ligand; this stabilization facilitates penetration of tumor cell membrane and causes the destruction of tumor cell [32].

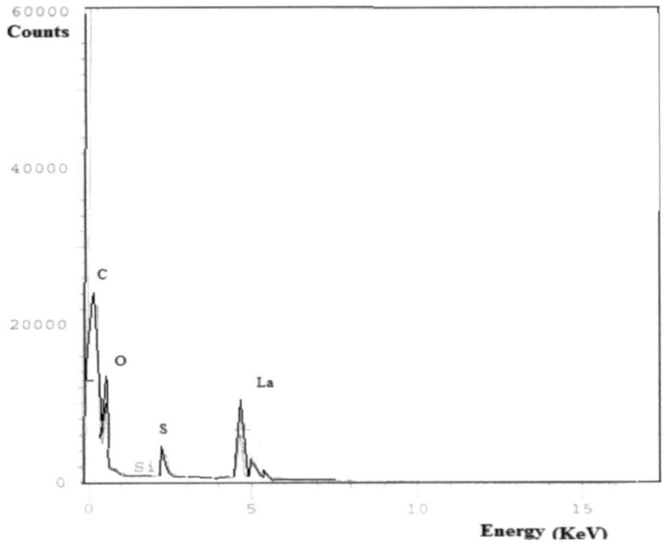

Figure 3.
EDX spectra of nano-La(III) complex 2.

Figure 4.
SEM images of nanoparticles as produced by ultrasound [(La)(L¹)(NO₃)(H₂O)] (1).

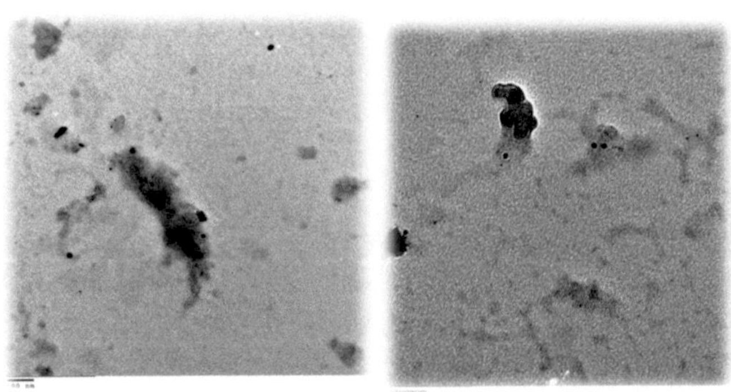

Figure 5.
TEM image of La(III) nanocomplex 2.

3.11 TEM analysis

Transmission electron microscopy (TEM) can be used to directly image nanoparticles at scales approaching a single atom. TEM analysis is performed to examine the size and shape of the nanoparticles. The La(III) complex **2** nanoparticle was fairly uniform in size, spherical in shape, and with the average diameter

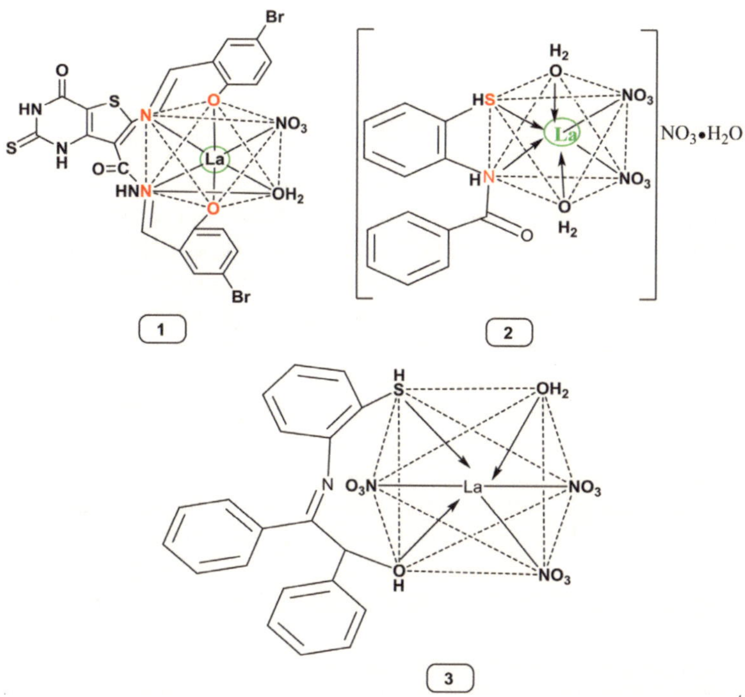

Figurc 6.
Suggested structure of metal complexes **1–3**.

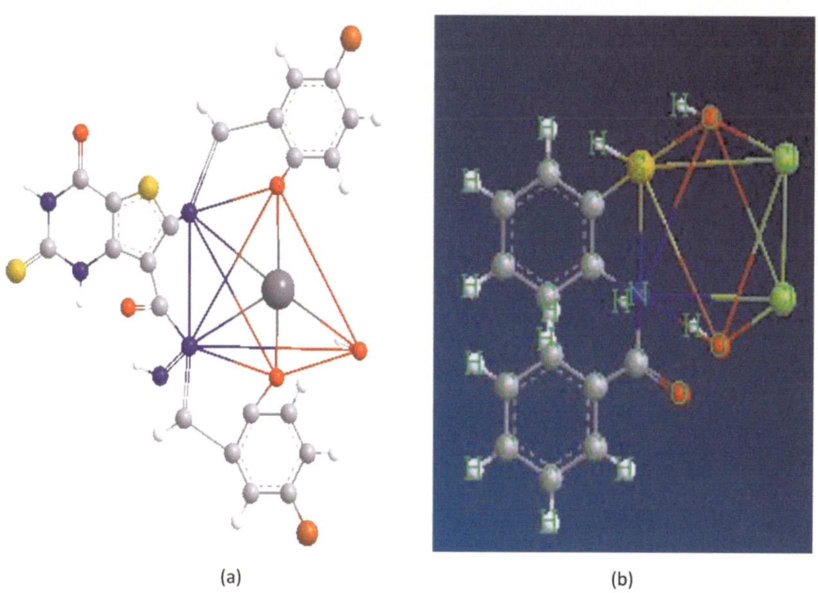

Figure 7.
Geometry-optimized structures of (a) complex **1** *(b) La complex* **2**.

ranging from 3.2 to 44.9 nm (**Figure 5**). Electron microscopy analysis allowed confirming visually the observed stability of the obtained nanocomplex.

Correlation of all results obtained for the complexes under study gives us information regarding the suggested structure of the complexes to be as in **Figure 6**. The fully optimized geometries of the ligand and its metal complexes were shown in **Figure 7**.

4. Title

4.1 Cytotoxicity and anticancer activity

The cytotoxicity of the prepared nanocomplexes has been demonstrated and tested for their anticancer activity against Caco2, Vero, and MCF-7 cell lines. The obtained results revealed that complex **1** had the best recorded anticancer activity against Caco2 cells with inhibition percentage of 72.13 (**Figure 8B**). The IC_{50} (μg) (the value which corresponds to the concentration required for 50% inhibition cell viability) for the noncytotoxic doses against human primary dermal fibroblasts cells were recorded as 2.482 μg/ml. The IC_{50} for these structures were recorded as 0.3772 μg/ml for compound **1** on Caco2 cells. In spite of the potency of the sample as anticancer agents, the superior selectivity index values against Caco2 cells than noncancerous cells were recorded with a value of 46.6.

On the other hand, the other tested compounds have shown moderate activities against all cells; the anticancer activity is less than that recorded for **1**.

4.2 Flow cytometry

Flow cytometric analyses of propidium iodide-stained nuclei cells were performed to investigate whether treatment could induce cell cycle perturbations in human epithelial colorectal adenocarcinoma. Cell cycle parameters were compared for Caco2 cells that had been incubated for 24 h with 1 ml of IC_{50} concentrations of sample **1** with untreated (control) cells. As shown in **Figure 9**, following 24 h incubation with treatment, there was an increase in the proportion of cells in the G2/M phase (51.10%, for treatment **1**).

Also, an additional peak was observed, consistent with aneuploid cells which contain more than 4n DNA. This peak may represent a population of cells that has escaped mitotic arrest and continued to replicate as multinuclear cells without dividing (i.e., endoreduplication), or it may be due to a small fraction of tetraploid

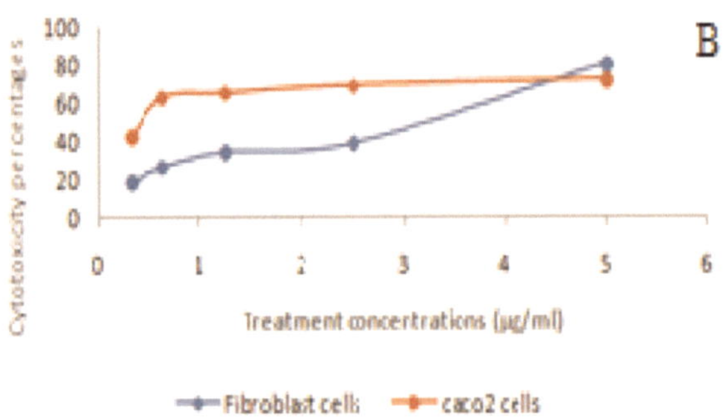

Figure 8.
Cytotoxicity and anticancer activity of the prepared compounds against human primary dermal fibroblasts and Caco2 cell lines, respectively.

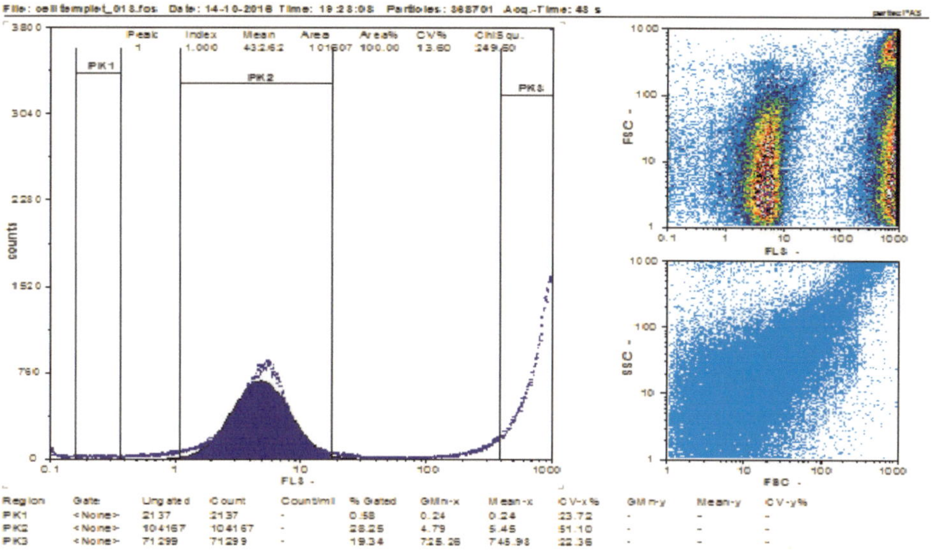

Figure 9.
Flow cytometric analysis of compound 1, against Caco2 cell line.

cells which are already present in the culture, which undergo G2/M arrest, resulting in cells with 8n DNA.

4.3 Chitosan-nanocomplex 2 composite nanofibers

The chitosan played an important role in the synthesis of the composite nanofiber. The amino group has been used to stabilize metal nanoparticles [33], so we suggested that the free NH_2 ligands of chitosan might bind to the nanoparticles and stabilized them.

4.3.1 FTIR

The formation of the CS-stabilized NPs was further investigated by FTIR spectroscopy. The FTIR spectra of the CS-NP nanofiber and CS nanofiber attributed a broad band at the range of 3350–3510 cm^{-1} corresponding to the amine and hydroxyl groups. For the nanocomposite, there is a change in the shape and frequencies of the bands in this region, indicating participation of the NH_2 group in the reaction. Although there is some difference in the two figures, both of them showed the basic characteristic peaks for CS at 3422 cm^{-1} (O─H stretch) and 2960–2970 cm^{-1} (C─H stretch), which indicated that the CS can adsorb on the formed nanoparticles in the preparation process [34, 35].

4.3.2 TGA

Thermogravimetric analysis (TGA) is widely used to investigate the thermal decomposition of a polymer. The TGA thermograms of pristine CS nanofiber, CS-NP composite nanofiber, at a heating rate of 20°C/min under nitrogen atmosphere are investigated. Chitosan's nanofiber weight change between 28.27 and 100°C is associated with the loss of adsorbed and bound water. Chitosan polymer degradation started at 100°C and continued up to 333.80°C, with a 75.5% polymer weight loss [36, 37].

The CS-NP composite nanofiber exhibited a three-step degradation pattern. The first step occurred between 24.68°C and 100.86°C, which is due to water evaporation. The second and the major mass loss were observed between 100°C and 261.34°C, with a 40.42% polymer weight loss. The third step occurred between 261.34 and 367.43°C, with a 19.95% polymer weight loss with total loss of 71.27% at 367.43°C.

4.3.3 TEM

The NP complex was encapsulated by CS, and their sizes were in the range of 15–50 nm as shown in the TEM image of the CS–NP composite nanofiber (**Figure 10**). It could be observed that a considerable amount of non-spherical particle was observed. These non-spherical particles seemed to be fabricated by coalescence of two spherical particles during the nucleation process [38].

4.3.4 SEM

The morphology of electrospun (chitosan–nanoparticle) composite nanofiber is shown in **Figure 11**. It is clear from these figures that no phase separation has been observed and the homogeneity of the obtained nanofibers can be easily observed and the composite of the nanoparticles has been proven. It is also clear that the diameter of chitosan-nanoparticle composite nanofiber is smaller than chitosan nanofiber.

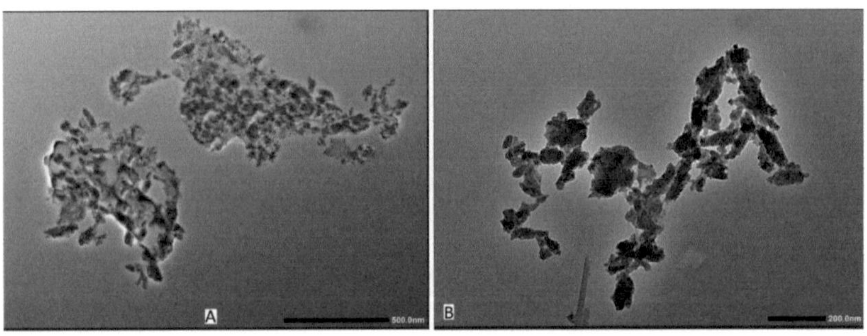

Figure 10.
TEM micrograph of (A) chitosan nanofiber and (B) (chitosan-nanoparticle) composite nanofiber.

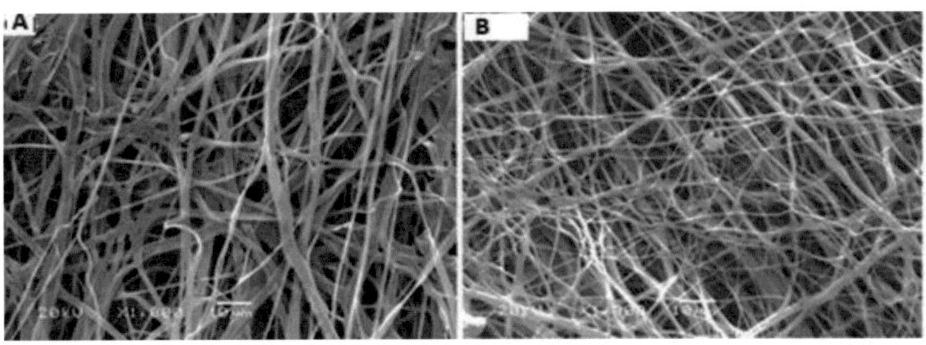

Figure 11.
SEM micrograph 1000× of (A) CS nanofiber and (B) La-CS composite nanofiber.

4.3.5 Antimicrobial activity

The ability of prepared La nanoparticle-coated nanofibers to cease the growth of microbes known for its human pathogenicity has been investigated. La nanoparticle-coated nanofibers succeeded to locally cease the growth of all tested microbes. Each disc succeeded to stop the growth of tested microbes at the boundaries between the disc and the plates' surfaces, forming clear zones with 7 mm diameter. It is worth mentioning that La nanoparticle-coated nanofibers showed the same pattern of antimicrobial activity against all tested microbes. The ability of prepared nanoparticle-nanofiber hybrid to affect the microbes locally is definitely promising for biomedical applications. These results would support the coexistence of the microbes and their antimicrobial agent at the same location and hence improve the overall microbe drug interaction process. Moreover, this kind of antimicrobial agent localization would prevent the probability of its migration to other sites inside the body that would help to avoid the expected subsidiary complications. It would be recommended that these nanoparticle-coated nanofibers are used as wound dressing for skin injuries.

5. Conclusion

La(III) nanocomplexes with different organic ligands were synthesized and characterized by different techniques. In addition to the nanosize was proved by different techniques also. All prepared metal complexes were screened against different human cell lines. It was found that we were able to identify new potent and selective anticancer compounds. Also, one of these nanocomplexes was tested as wound dressing which is recommended to be used as local chemotherapy for skin injuries.

Author details

Rania H. Taha[1,2]

1 Chemistry Department, College of Science, Jouf University, Sakaka, Saudi Arabia

2 Department of Chemistry, Faculty of Science (Girls), Al-Azhar University, Nasr City, Cairo, Egypt

*Address all correspondence to: mhmd_mosad@yahoo.com; rania@azhar.edu.eg

References

[1] Dunn PJ. The importance of green chemistry in process research and development. Chemical Society Reviews. 2012;**41**:1452-1461

[2] Gou Y, Li J, Fan B, Xu B, Zhou M, Yang F. Structure and biological properties of mixed-ligand Cu(II) Schiff base complexes as potential anticancer agents. European Journal of Medicinal Chemistry. 2017;**134**:207-217

[3] Binil PS, Anoop MR, Jisha KR, Suma S, Sudarsanakumar MR. Synthesis, spectral characterization, thermal and biological studies of lanthanide(III) complexes of oxyphenbutazone. Journal of Rare Earths. 2014;**32**:43

[4] Bouchoucha A, Terbouche A, Bourouina A, Djebbar S. New complexes of manganese (II), nickel (II) and copper (II) with derived benzoxazole ligands: Synthesis, characterization, DFT, antimicrobial activity, acute and subacute toxicity. Inorganica Chimica Acta. 2014;**418**:187

[5] Saad FA, El-Metwaly NM, Farghaly TA, Elghalban MG, Shah RK, Al-Hazmid GA, et al. Illustration for series of new metal ion complexes extracted from pyrazolone derivative, spectral, thermal, QSAR, DFT/B3LYP, docking and antitumor investigations. Journal of Molecular Liquids. 2017;**229**:614

[6] Delgado S, Santana A, Castillo O, Zamoro F. Dynamic combinatorial chemistry in a solvothermal process of Cu(I,II) and organosulfur ligands. Dalton Transactions. 2010;**39**:2280

[7] Hsu H-Y, Tseng C-C, Chung-Ming Sun BM. Ionic liquid-supported synthesis of dihydroquinazolines and tetrahydroquinazolines under microwave irradiation. Molecular Diversity. 2012;**16**:241

[8] Baghbanzadeh M, Pilger C, Kappe CO. Rapid nickel-catalyzed Suzuki-miyaura cross-couplings of aryl carbamates and sulfamates utilizing microwave heating. The Journal of Organic Chemistry. 2011;**76**:1507

[9] Qin W, Long S, Panunzio M, Biondi S. Schiff bases: A short survey on an evergreen chemistry tool. Molecules. 2013;**18**:12264

[10] Manju N, Mishra D, Kumar R. Coordination chemistry of Schiff base tin complexes. Journal of Coordination Chemistry. 2014;**40**:343

[11] Evans CH. Biochemistry of the Lanthanides, Lanthanides Series Determination by Various Analytical Methods. New York & London: Plenum Press; 1990

[12] Biba F, Groessl M, Egger A, Roller A, Hartinger CG, Keppler BK. New insights into the chemistry of the antineoplastic lanthanum complex tris (1,10-phenanthroline)tris (thiocyanato-κ*N*)lanthanum(III) (KP772) and its interaction with biomolecules. European Journal of Inorganic Chemistry. 2009; **2009**:4282

[13] Kostova I, Momekov G, Zaharieva M, Karaivanova M. Cytotoxic activity of new lanthanum (III) complexes of bis-coumarins. European Journal of Medicinal Chemistry. 2005;**40**:542

[14] Kostova I, Momekov G. New cerium (III) complexes of coumarins—Synthesis, characterization and cytotoxicity evaluation. European Journal of Medicinal Chemistry. 2008; **43**:178

[15] Kulkarni A, Patil SA, Badami PS. Synthesis, characterization, DNA cleavage and in vitro antimicrobial studies of La(III), Th(IV) and VO(IV)

complexes with Schiff bases of coumarin derivatives. European Journal of Medicinal Chemistry. 2009;**44**:2904

[16] Chen JG, Qiao X, Gao YC, et al. Synthesis, DNA binding, photo-induced DNA cleavage and cell cytotoxicity studies of a family of light rare earth complexes. Journal of Inorganic Biochemistry. 2012;**109**:90

[17] Mogosanu GD, Grumezescu AM. Natural and synthetic polymers for wounds and burns dressing. International Journal of Pharmaceutics. 2014;**463**:127-136

[18] Sedghi R, Shaabani A, Mohammadi Z, Yazdi F, Isaei E. Carbohydrate Polymers. 2017;**159**:1-10

[19] Almahdy O, EL-Fakharany EM, EL-Dabaa E, Ng TB, Redwan EM. Examination of the activity of camel milk casein against hepatitis C virus (genotype-4a) and its apoptotic potential in hepatoma and HeLa cell lines. Hepatitis Monthly. 2011;**11**: 724-730

[20] El-Fakharany EM, Sánchez L, Al-Mehdar HA, Redwan EM. Effectiveness of human, camel, bovine and sheep lactoferrin on the hepatitis C virus cellular infectivity: Comparison study. Virology Journal. 2013;**10**:199

[21] Mosmann T. Rapid colorimetric assay for cellular growth and survival: Application to proliferation and cytotoxicity assays. Journal of Immunological Methods. 1983;**65**:55-63

[22] Léonce S, Perez V, Lambel S, Peyroulan D, Tillequin F, Michel S, et al. Induction of cyclin E and inhibition of DNA synthesis by the novel acronycine derivative S23906-1 precede the irreversible arrest of tumor cells in S phase leading to apoptosis. Molecular Pharmacology. 2001;**60**:1383-1391

[23] Aly HM, Taha RH, El-Deeb NM, Alshehri A. Efficient procedure with new fused pyrimidinone derivatives, Schiff base ligand and its La and Gd complexes by green chemistry. Inorganic Chemistry Frontiers. 2018;**5**: 454-473

[24] Fasina MT, Ogundele OO, Ayeni I. Synthesis and biological properties of N_2O_2 Schiff bases derived from o-phenylenediamine and substituted salicylaldehydes. Journal of Chemical and Pharmaceutical Research. 2014;**6**: 816-819

[25] Gomathi V, Selvameena R. Synthesis, spectroscopic, electrochemical and biological studies of Schiff base complexes of some 3d transition metals derived from 2-aminophenol and 2-hydroxynaphthaldehyde. Indian Journal of Applied Research. 2013;**3**(4):51-53

[26] Rania HT. Preparation. Spectroscopic Characterization, DNA Cleavage, Antimicrobial and Antitumor Investigations of Nickel and Uranyl Schiff Base Complexes in Bulk and Nano size. Current Science International. 2015; **4**(4):684-700

[27] Khandelwal A, Agrawal M, Lamba S, Baswal G. Synthesis and characterization of mixed ligand complexes of Mn(III) with 5-bromosalicylaldehyde and 2-hydroxyaryl carbonyl compounds or Âdiketones. Weekly Science Research Journal. 2013;**1**:1-10

[28] Rania HT, Zienab AE, Aida AS, Esmail ME, Mai MM. Synthesis and characterization of newly synthesized Schiff base ligand and its metal complexes as potent anticancer. Journal of Molecular Structure. 2019;**1181**:536-545

[29] Mohebbi S, Bakhshi B. Electrochemical and spectral behavior

of mononuclear oxovanadium(IV) salicyldiimine complexes. Journal of Coordination Chemistry. 2008;**61**:2615

[30] Podorov SG, Faleev NN, Pavlov KM, Paganin DM, Stepanov SA, Forster E. New approach to wide-angle dynamical X-ray diffraction by deformed crystals. Journal of Applied Crystallography. 2006;**39**:652

[31] Soheir SA, Rania HT, Amany MT, Mohamed OA, Hanady AM. Antimicrobial activity of bio and chemical synthesized cadmium sulfide nanoparticles. The Egyptian Journal of Hospital Medicine. 2018;**70**(9):1494-1507

[32] Mandegani Z, Mozaffar A, Zahra A, Afshan M, Nasser I, Akbar O. A nano tetraimine Pd(0) complex: Synthesis, characterization, computational studies and catalytic applications in the Heck–Mizoroki reaction in water. Green Chemistry. 2015;**17**:3326-3337

[33] Selvakannan PR, Mandal S, Phadtare S, Pasricha R, Sastry M. Langmuir. 2003;**19**:3545-3549

[34] Kumar CV, McLendon GL. Chemistry of Materials. 1997;**9**:863-870

[35] He P, Hu N, Rusling JF. Langmuir. 2004;**20**:722-729

[36] Rosi M, IskandarAbdullah F, Khairurrijal M. Hydrogel-polymerelectrolytes based on polyvinyl alcohol and hydroxyethylcellulose for supercapacitor applications. International Journal of Electrochemical Science. 2014;**9**(8):4251-4256

[37] Abu-Saied MA, Wycisk R, Abbassy MM, El-Naim GA, El-Demerdash F, Youssef ME, et al. Sulfated chitosan/PVA absorbent membrane for removal of copper and nickel ions from aqueous solutions—Fabrication and sorption studies. Carbohydrate Polymers. 2017; **165**(1):149-158

[38] Bharathi S, Fishelson N, Lev O. Langmuir. 1999;**15**:1929-1937

[39] Abu-Saied MA, Khalil KA, Al-Deyab SS. Preparation and characterization of poly vinyl acetate nanofiber doping copper metal. International Journal of Electrochemical Science. 2012;**7**: 2019-2027

Chapter 3

Infrared Irradiation, an Excellent, Alternative Green Energy Source

Joel Martínez and René Miranda

Abstract

In this chapter, attending the request by the UNESCO, *Decade to Educate in the Sustainability*, a green-chemical approach is offered. Thus, the most recent advances using the infrared irradiation as alternative energy source are given; this is in order to activate an organic reaction in addition to conveniently perform the extractions of secondary metabolites. It is worth noting that this manuscript is an improved extent of a review previously reported by our group. The chapter is accomplished considering two sections, displaying a comprehensive overview: in a first instance, after a deep search in the literature, a broad compilation of suitable information is presented; in a second landscape, recent unpublished results from our and other laboratories are described.

Keywords: infrared irradiation, green chemistry, UNESCO-DES, sustainable development, reaction activation, natural product extraction

1. Introduction

Since the born (early 1990s) of the paradigm of the green chemistry (GC), it has acquired its up-to-date position as a scientific discipline preventing pollution and consequently contributing to the sustainable development (SD) [1]. In this sense, it is important to keep in mind the request of the United Nations Educational, Scientific and Cultural Organization (UNESCO) performed by the establishment of the Decade to Educate in the Sustainability (DES), 2005–2014, satisfactorily evaluated in November 2014 at Aichi, Nagoya, Japan. In this sense, the UNESCO desire to improve the quality of education (teaching researching and informing) at all levels, transforming the society by the reorientation of the education adopting the sustainable development aims [2, 3].

Consequently, many educational institutions in all levels, around the world, have assumed the GC protocol as an efficient pathway to contribute within the SD using and implementing innovative technologies.

In recent years, scientists around the world have implemented strategies to avoid the use of fossil resources; in this regard, it is convenient to note that between the GC-protocol, the six principles (*energy requirements should be more recognized for their environmental and economic impacts, and they must be minimized*) [4, 5] are well recognized [6] being important to highlight that several approaches to reduce the energetic consumption have been developed in order to diminish the dependence of fossil resources: the use of microwave irradiations (MW), ultrasound (US), mechanical milling (MM), photochemical (PC), electrochemical (EC), all of them

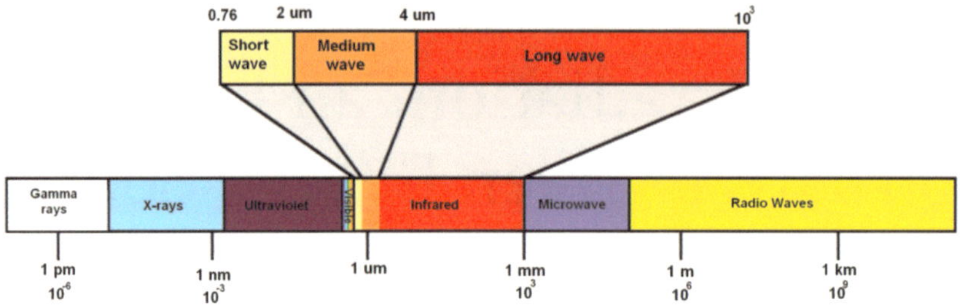

Figure 1.
Infrared irradiation zones.

well recognized by the GC [7]. More recently, infrared irradiation (IR) has emerged as another excellent activating source into the green chemistry efforts [1, 7].

The infrared irradiation has been already perceived by the scientific community for the convenient promotion of organic transformations, this in addition to its employment to perform efficient natural product extractions, among others process.

Regarding the last commentaries, it results appropriated to establish the goal of this chapter: provide after a profound search in the literature, the actual status concerned to the employment of the infrared irradiation as an alternative mode to activate chemical reactions in addition to demonstrate its invaluable use to perform conveniently the extraction of interesting secondary metabolites.

On the other hand, taking as supportive reference the electromagnetic spectrum, the infrared region is divided into three zones, **Figure 1**. The shortwave or near-infrared zone has band spans from 0.76 to 2 μm (NIR); the middle or medium infrared zone, with band spans from 2 to 4 μm (MIR); and the zone of long wave or far infrared with band spans from 4 to 1000 μm (FIR).

It is worth noting that the NIR offers important advantages to carry out a reaction: due to its immediate response time, because it is efficiently used by convection, in addition to the longer life time of the tungsten-halogen filament [1, 8].

2. Infrared irradiation, actual use around the world

Considering the use of infrared irradiation, as an activating mode of reaction in complement to its employment to activate the extraction of secondary metabolites, in this chapter after a profound literature search, the obtained information was organized and presented in several sections. Each one of these sections was performed in order to highlight the actual status of the use of the infrared irradiation for the purposes previously commented. It is also important to note that the manuscript was created bearing in mind to attend the Decade to Educate in sustainable Development involving the green chemistry protocol.

Therefore, in the first place, the upgraded use of infrared irradiation around the world including both the early and the most recent studies in this field is showed.

2.1 Knoevenagel condensation

In **Table 1**, three aldol type reactions are shown, employing MIR as the activating mode, using a natural bentonitic clay—Tonsil Actisil FF (TAFF) [9] as a catalyst. The target molecules were produced with good yields, short reaction times, in the absence of solvent, and involving easiest workup.

E	Reaction	Ref.
1	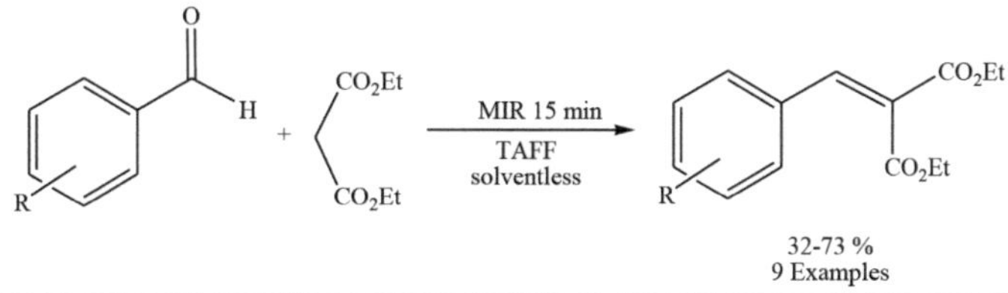	[10]
2	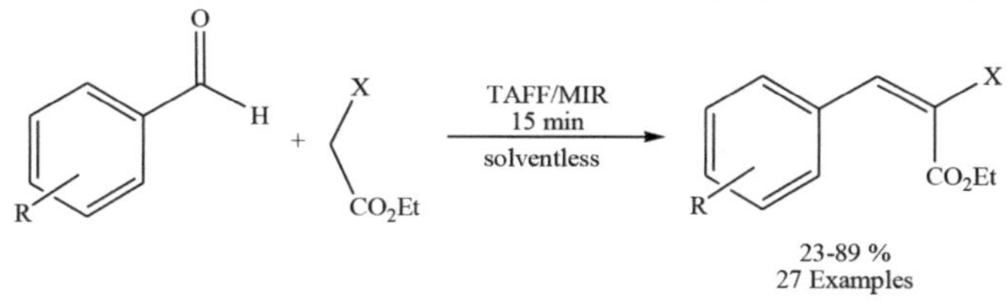	[11]
3	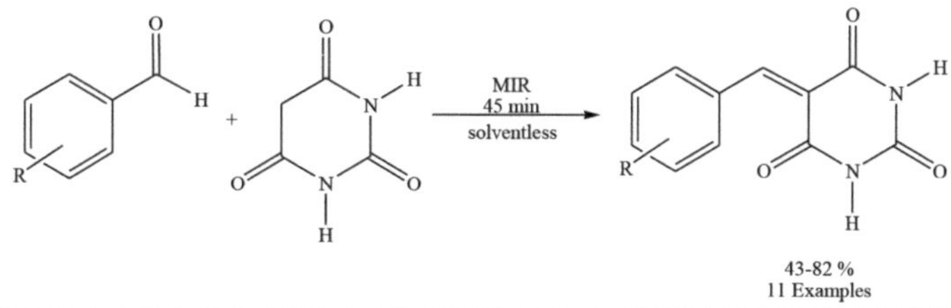	[12]

E = Entry, Ref = Reference.

Table 1.
Knoevenagel condensation reactions.

2.2 Nucleophilic additions

Several nucleophilic additions to various carbonylic substrates are confined in **Table 2**; consequently different products were obtained: heterocycles, Schiff bases, and the addition of a set of sulfur derivatives or the indolyl moiety to perezone, a natural product and its synthetic analog, Entry 10; it is important to note that the obtained indolyl molecules showed good activity against breast cancer cells.

In these processes, the MIR was employed to promote the reaction; in some cases, Co doped with ZnS nanoparticles or TAFF were used as catalyst. The reactions, in general, proceed with good to excellent yields. Also, a green procedure to obtain a set of five coumarins, was developed, comparing different activating modes, MW, NIR, US, MM, and mantle heating (MH), in the presence or absence of ethanol, Entry 9.

2.3 Related macrocycles-benzyl oligomers

Various related macrocycles or benzyl oligomers were produced in the presence of TAFF as catalyst, in absence of solvent, considering a comparative study between MIR and MW, both processes show similar yields, **Table 3**.

2.4 Multicomponent reaction

The multicomponent reaction implicates a very important and versatile strategy to generate an enormously number of reaction protocols to produce important molecules. In this sense, MIR and NIR have been usefully employed as activating reaction modes, to promote molecules related to the multicomponent protocols of

E	Reaction	Ref.
1	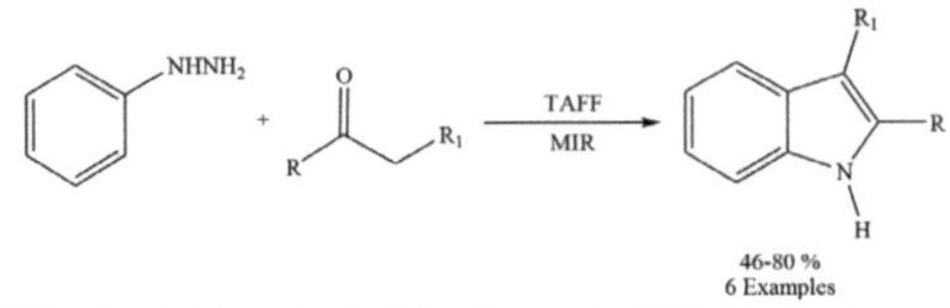	[13]
2	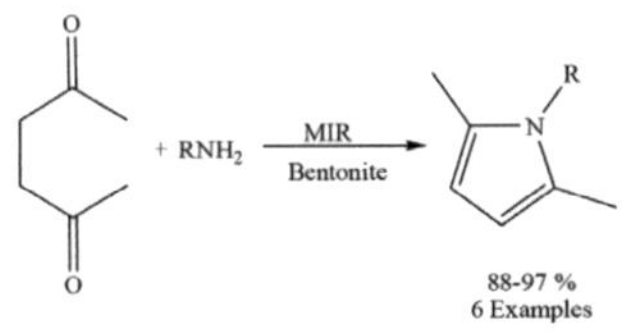	[14]
3	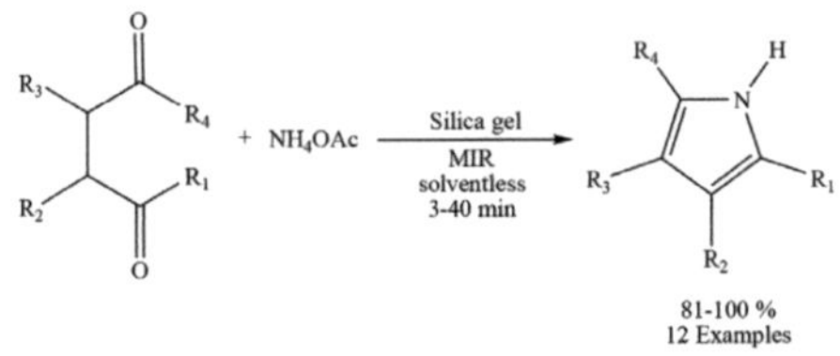	[15]
4	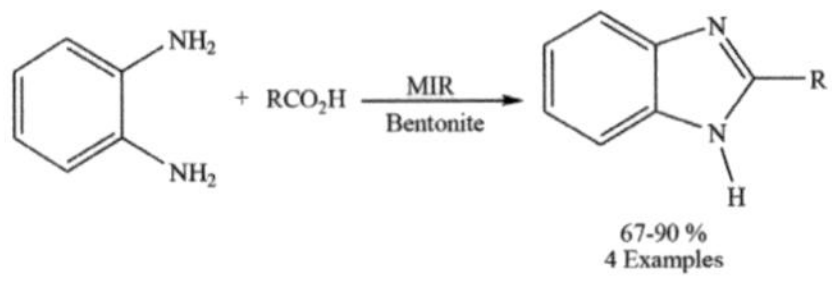	[16]
5	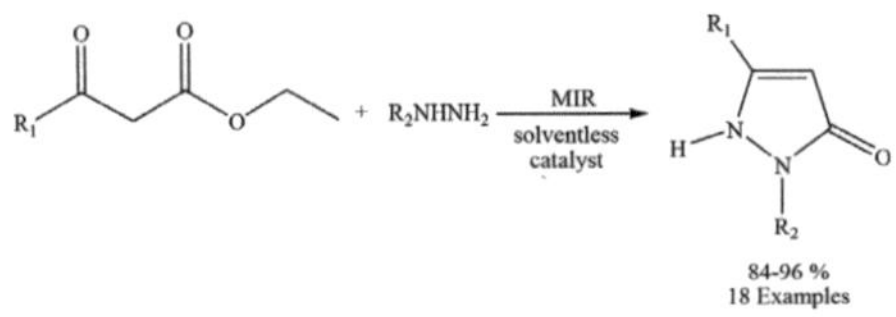	[17]
6	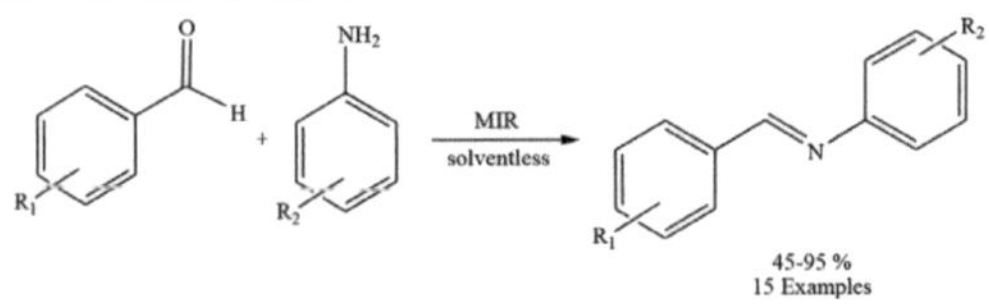	[18, 19]
7	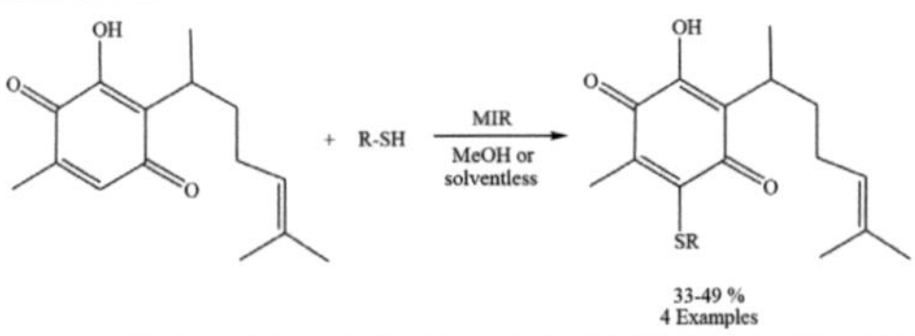	[20]
8	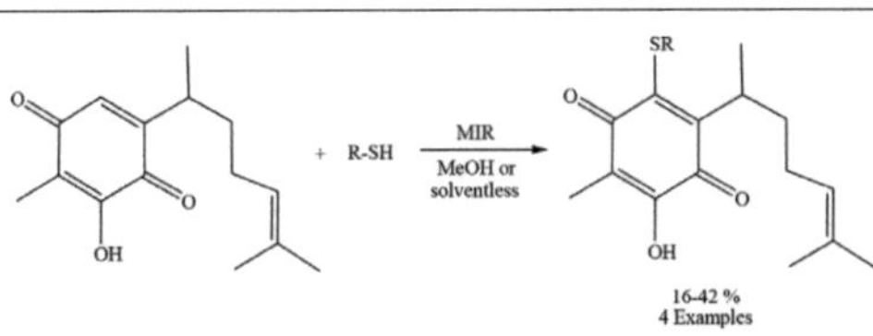	[20]
9	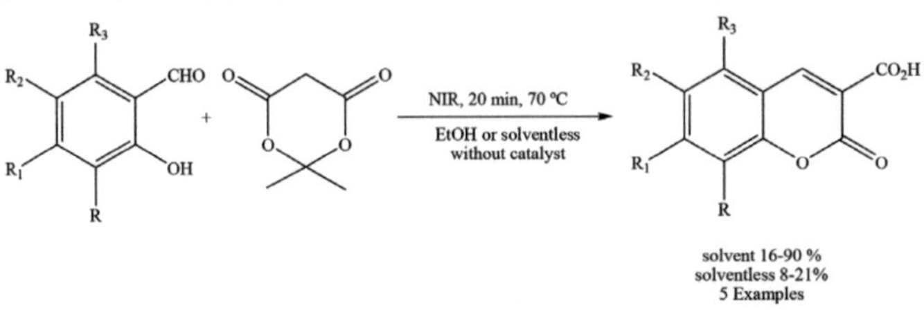	[21]

E	Reaction	Ref.
10	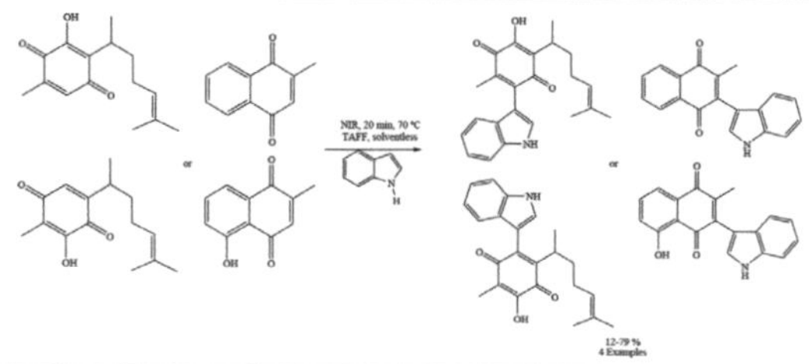	[22]

E = Entry, Ref = Reference.

Table 2.
Nucleophilic addition reaction.

E	Reaction	Ref.
1	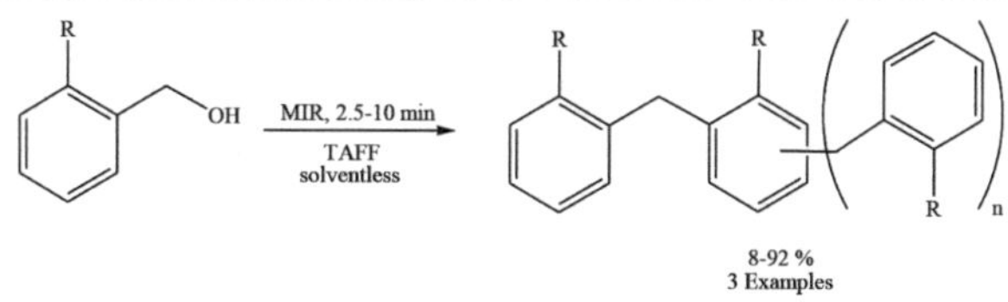	[23]
2	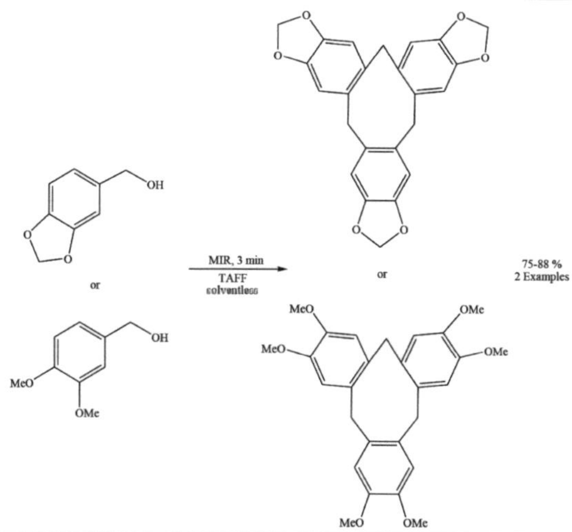	[23]

E = Entry, Ref = Reference.

Table 3.
Formation of related benzylic oligomers and macrocycles.

Biginelli, Hantzsch Meldrum, and Kamal-Qureshi (**Table 4**). The target molecules were in general with convenient green approach: without solvent or the use of EtOH as solvent, employing a green catalyst. The reported procedures involved short reaction times and easy workup.

2.5 Redox procedures

Important practices, within organic chemistry, are the reduction-oxidation, procedures. Thus, into the green chemistry protocol, several attempts have been developed, particularly within photocatalysis field, for example Guo and coworkers present the photothermic reduction of graphene oxide employing MIR [38].

E	Reaction	Ref.
1	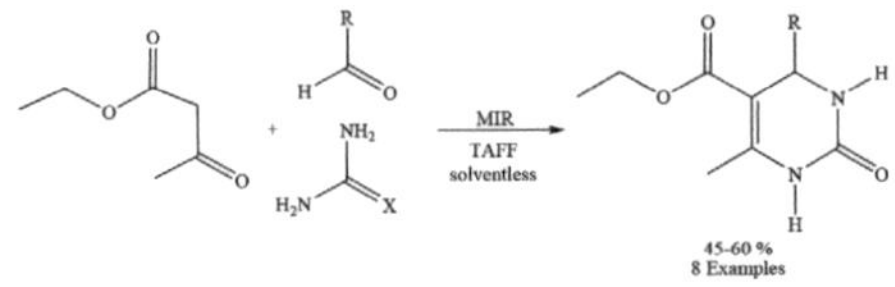	[24]
2	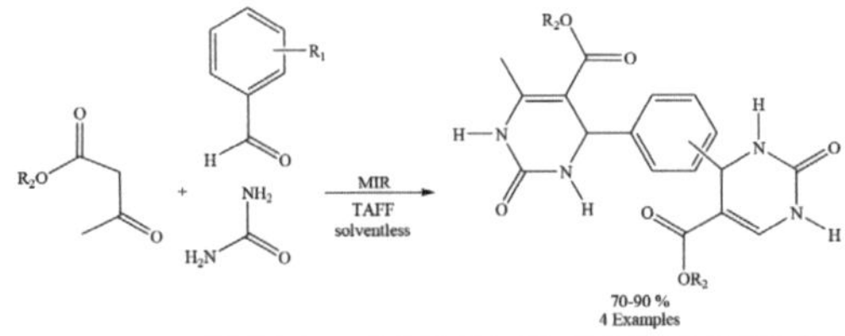	[24]
3	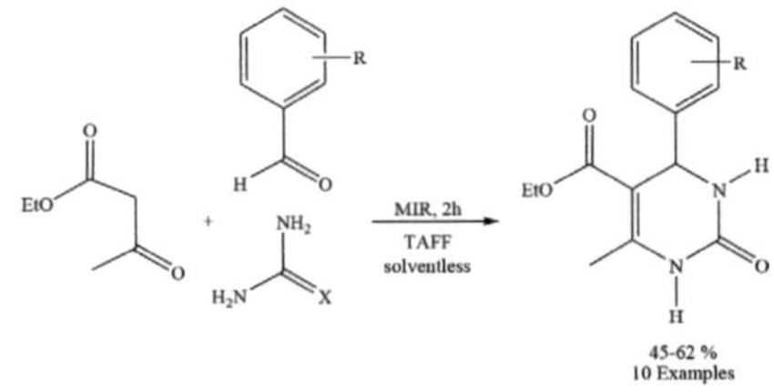	[25]
4	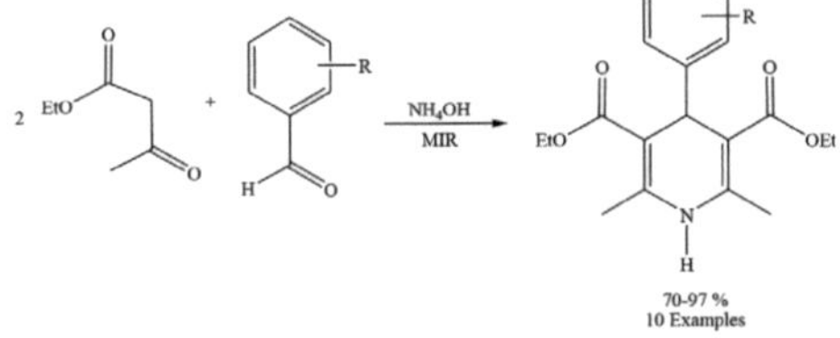	[26]
5	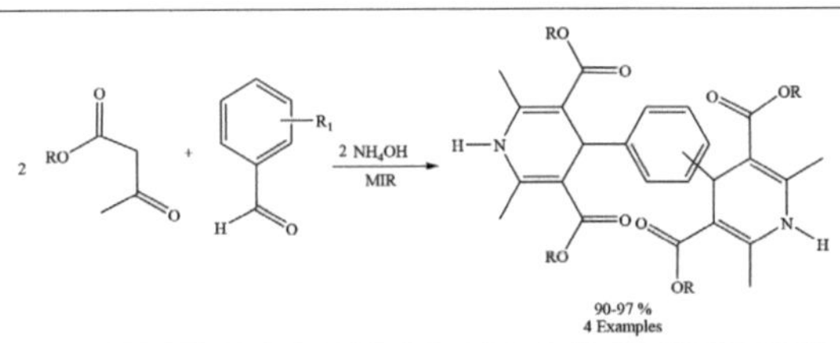 26	[26]
6	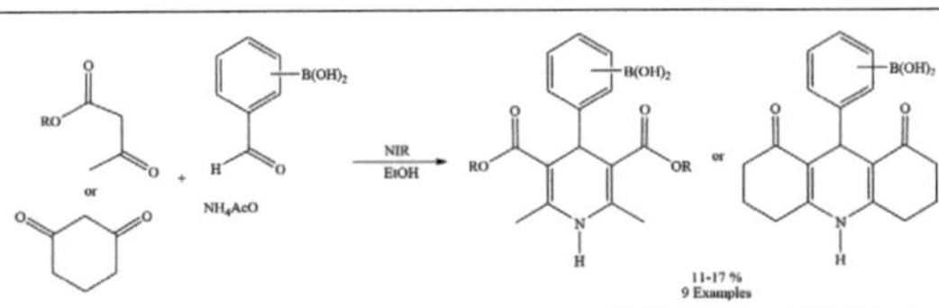	[27]
7	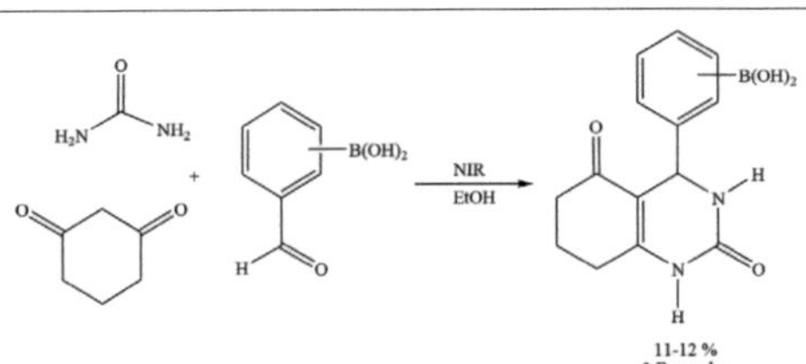	[27]
8	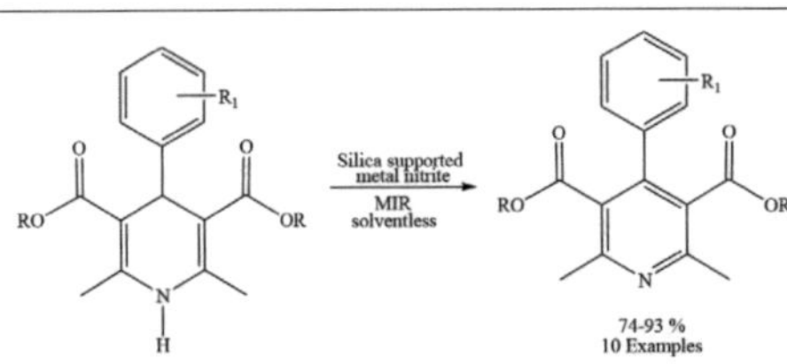	[28]

E	Reaction	Ref.
9	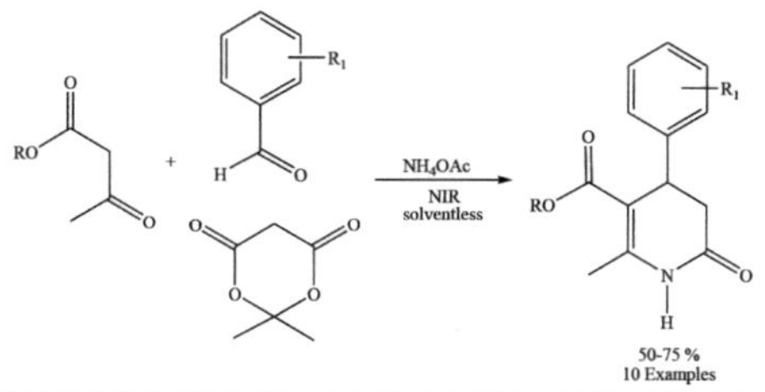	[29]
10	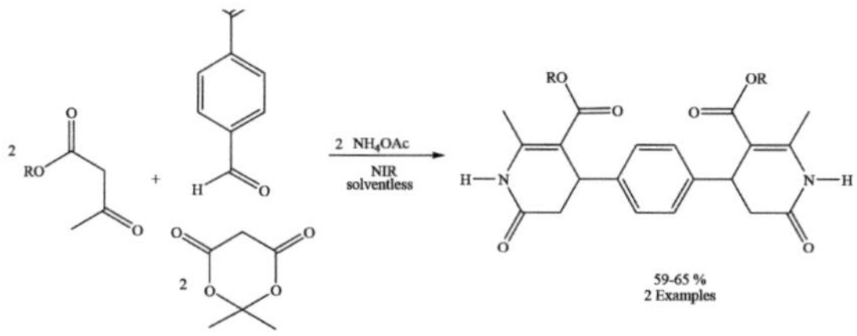	[29]
11	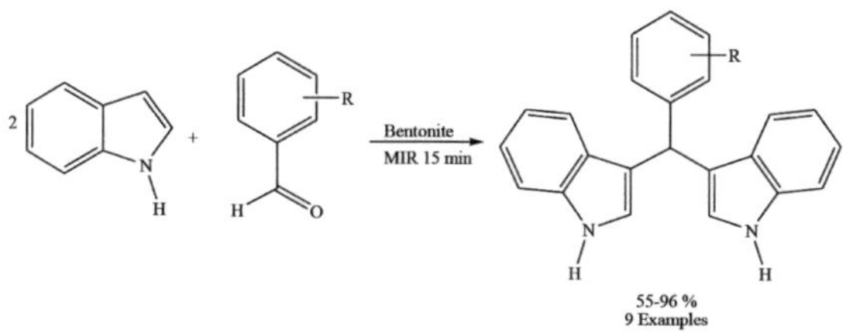	[30]
12	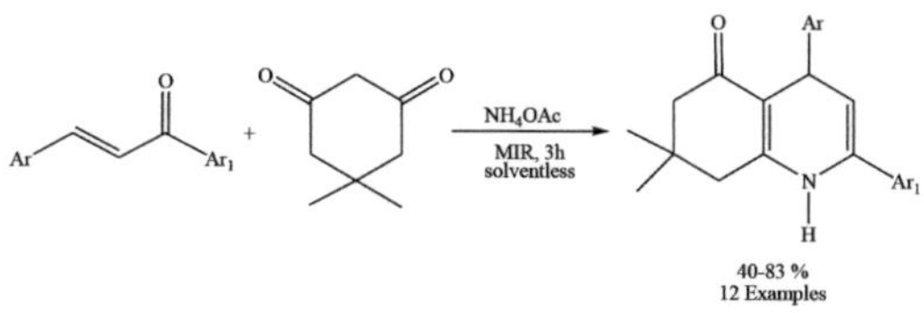	[31]
13	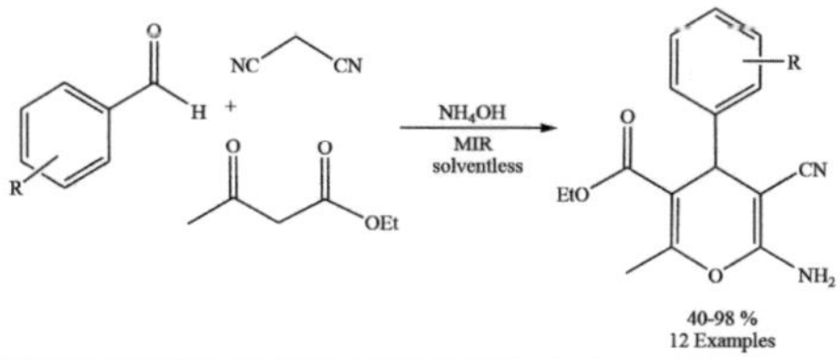	[32]
14	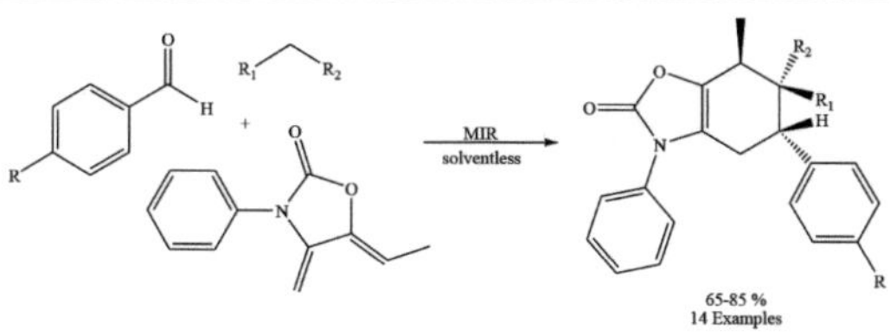	[33]
15	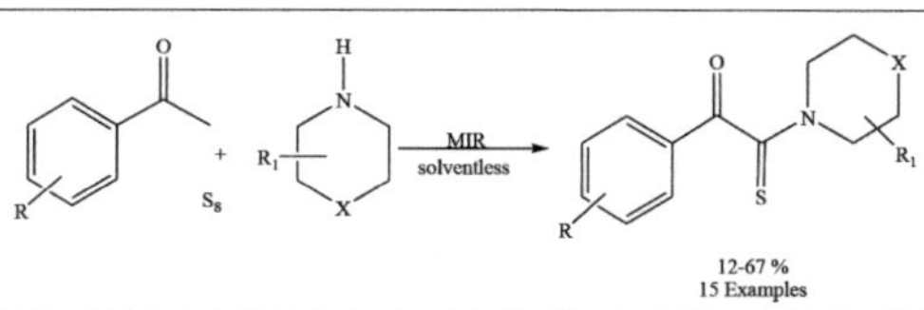	[34]
16	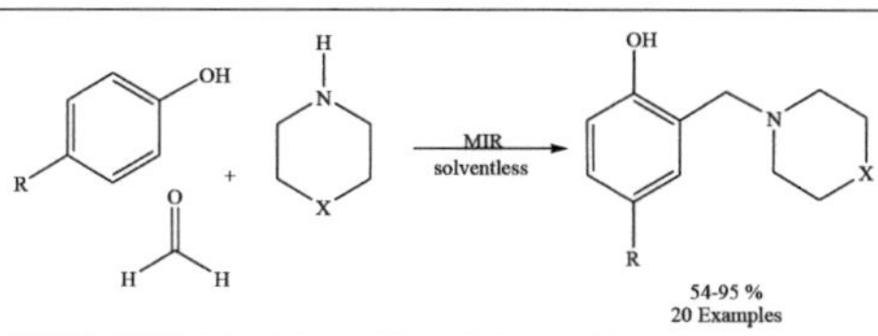	[35, 36]

E	Reaction	Ref.
17	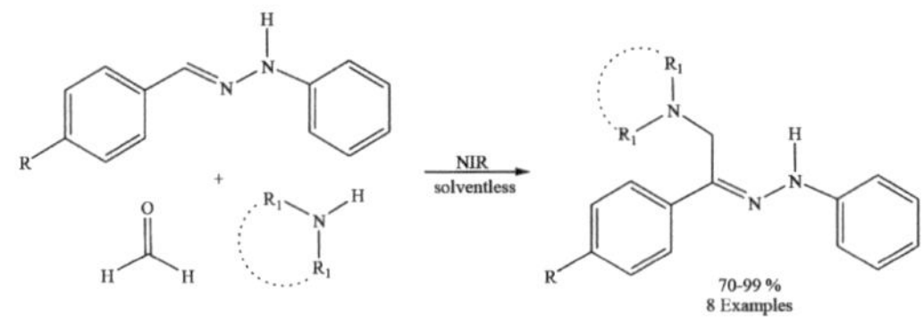	[37]

E = Entry, Ref = Reference.

Table 4.
Multicomponent reaction.

Another example, using NIR, is the photocatalytic decomposition of 2,4-dichlorophenol in aqueous solution with $Cu_2(OH)PO_4$ microcrystals [39], or the photo-oxidation of 1,3-diphenylisobenzofuran in aerated toluene [40]. Other oxidation processes are presented in **Table 5**, developed without solvent, excellent yields, and short time reaction.

2.6 Miscellaneous reactions

An interesting set of miscellaneous reactions is commented in this separated section. In this sense, the production of peptides by mean *N*-phosphoamino acid and amino acid in aqueous solution and irradiated with MIR was reported [43]. Also, the alpha-alumina powders were described by mean Pechini synthesis using MIR [44]. In other studies, SnO_2 nanoparticles were generated applying medium infrared irradiation [45]. The hexagonal form of boron nitride was employed to produce protective coatings using liquid polyborazylene as a boron source and was activated by MIR [46].

In a current research, the preparation of botanical samples enriched with organic matter by mean heating of focused far infrared light by dry ashing method is reported. In the dry process, the NIR was focused by reflection through the gold-plated layer; consequently, the energy efficiency was improved and a argon atmosphere was employed to avoid the loss of elements caused by sample combustion at high temperatures, 500°C, in short times of carbonization. It is important to note that this device shows several advantages as the instrument size is small compared

E	Reaction	Ref.
1	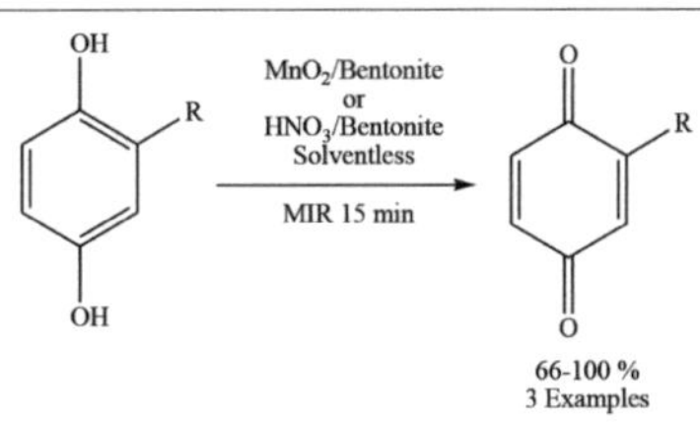	[41]
2	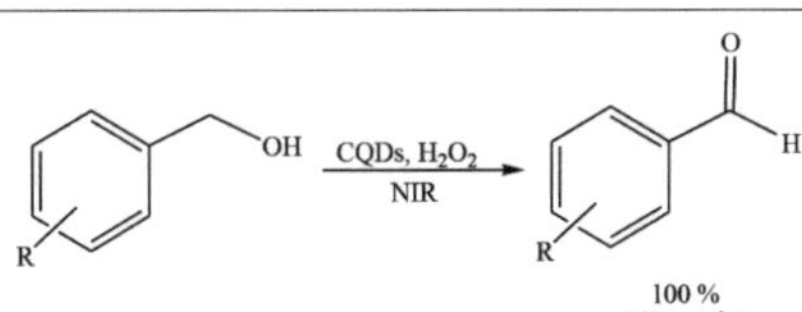	[42]

E = Entry, Ref = Reference.

Table 5.
Reduction-oxidation reactions.

with electric muffle furnace and has a fast heating rate the gold-plates improve the energetic efficiency; and in general, the ashing process is completed within 30 min, among others [47].

As a complement, in **Table 6**, several reactions are summarized as described in a previous report [1]. The included processes, in general involve short reaction times, good to excellent yields, the use of a catalyst and solventless conditions, and consequently a green approach is offered.

E	Reaction	Ref.
1	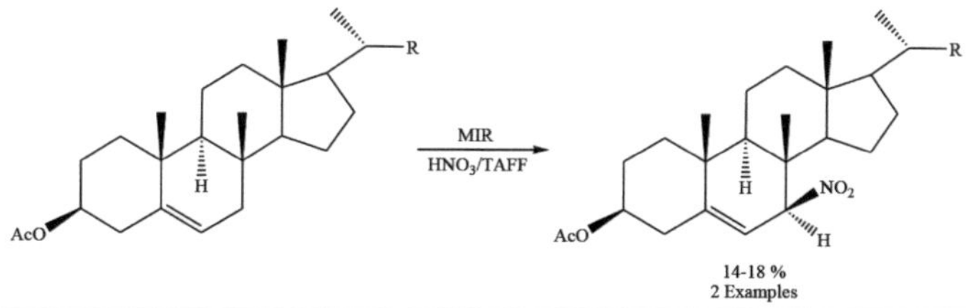	[48]
2	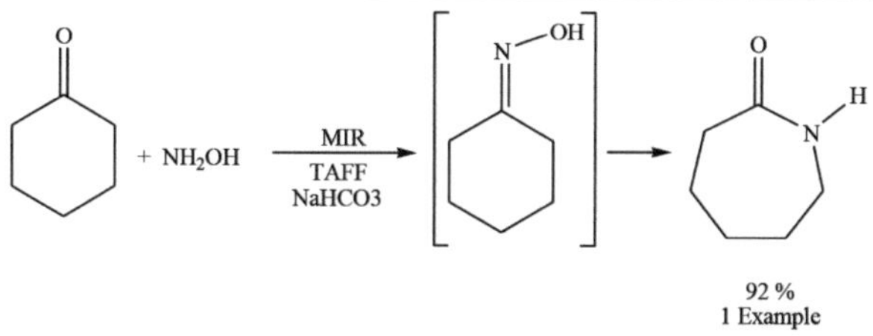	[49]
3	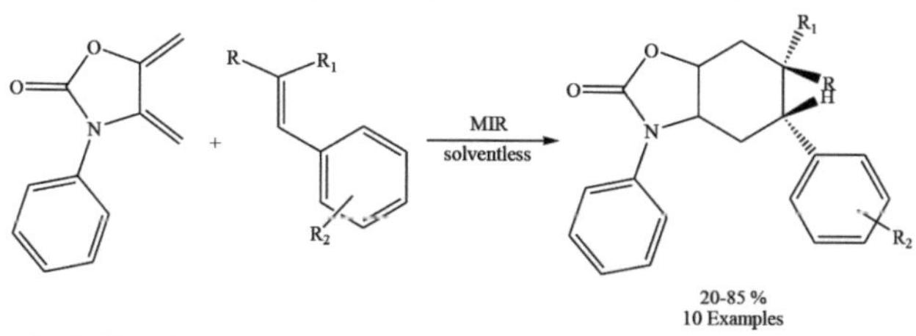	[33]
4	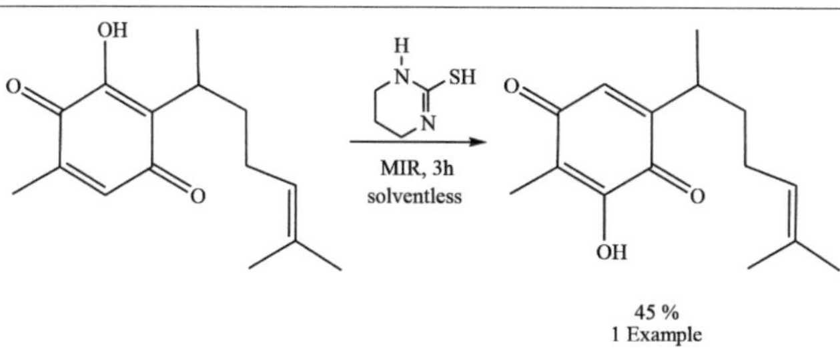	[20]
5	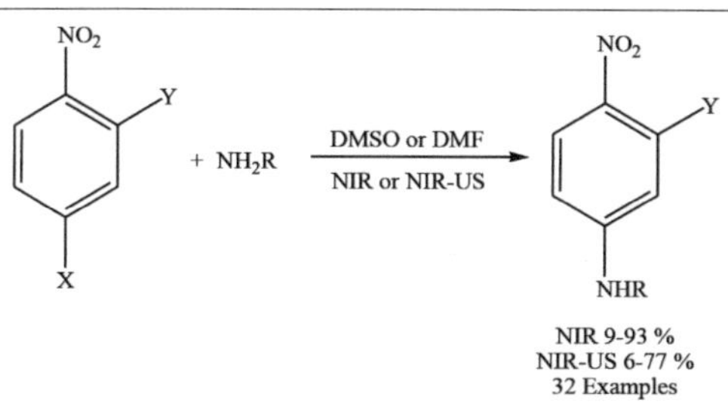	[50]

E = Entry, Ref = Reference.

Table 6.
Miscellaneous reactions.

Recently, several nucleophilic aromatic substitutions have been reported by Luna-Mora and coworkers; the authors inform about the preparation of *N*-(5-R^{1}amino-2-nitrophenyl) acetamides and 5-R^{1}amino-2-nitroanilines by means of a comparative study among MH, NIR, US, MW, and the combination of NIR-US here is concluded that the best process was when the NIR-US process is employed offering the best yields and short reaction times, Entry 5.

2.7 C–C bond formation

The C–C bond formation is a versatile tool to generate various compounds; some examples are presented in **Table 7**, the products were generated in good to excellent

E	Reaction	Ref.
1	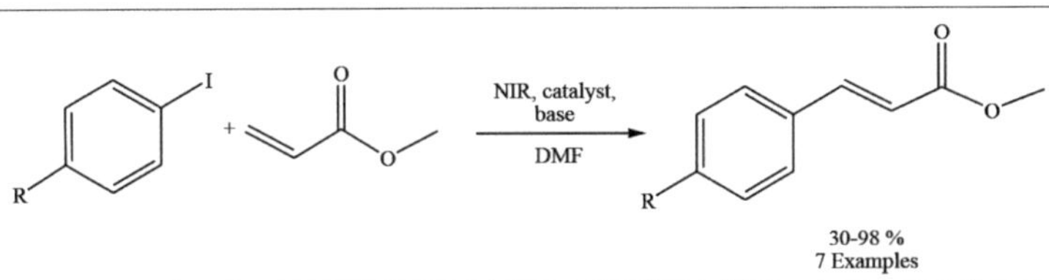	[37]
2	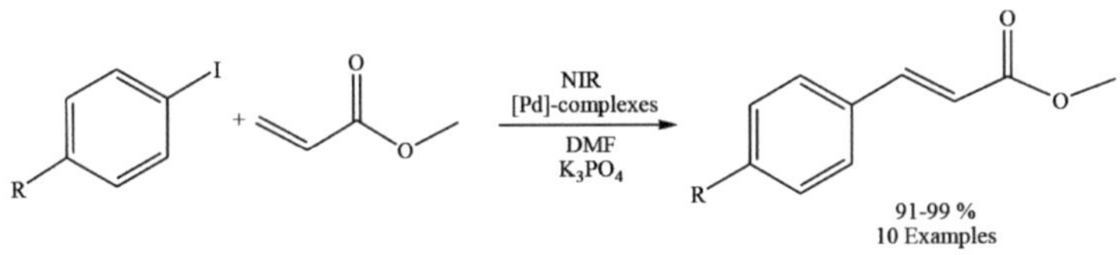	[51]
3	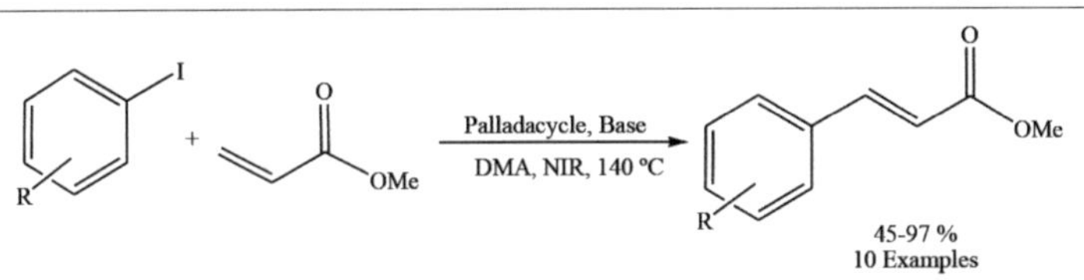	[52]
4	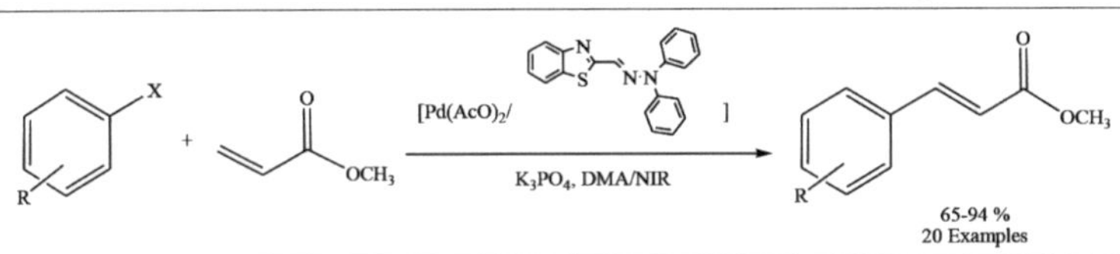	[53]
5	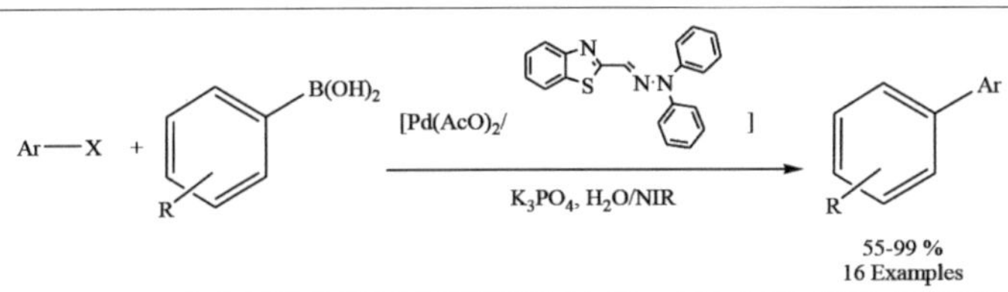	[53]
6	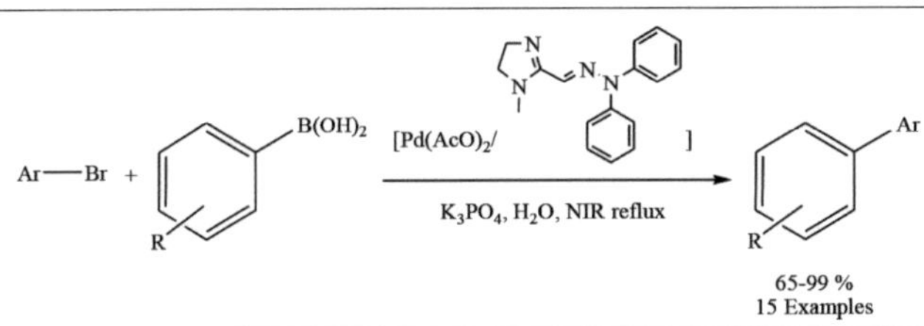	[54]

E = Entry, Ref = Reference.

Table 7.
Formation of C–C bond.

yields, using NIR as the activating mode of reaction, with short reaction times in the presence of a catalyst, with or without the presence of a base.

Related to **Table** 7, several interesting commentaries are performed:

Balam-Villareal in 2016 informed about the formation of C-C bond by means of a Mizoroki-Heck cross-coupling reaction, Entry 3, employing new sulfur-containing palladacycles as catalyst and aryl iodides with electron-donating and electron-withdrawing groups. Thus the NIR strategy, has similar behavior that the MH in relation to the yields obtained, but the time reaction was diminished by three times.

Another example, corresponding to C-C bond formation, implicates the use of arylhydrazones containing the benzothiazole moiety as a contribution to the Mizoroki-Heck reaction, employing palladium ligands, Entries 4 and 5. The authors report that the reaction was performed using near infrared irradiation. The products were obtained from good to excellent yields, reducing the reaction time. In the same report, it was reported the use of this arylhydrazones in the Suzuki-Miyaura cross-coupling reaction with NIR as alternative source, water as solvent, diminishing the time of reaction, and the products were obtained with good yields.

Continuing with the Suzuki-Miyaura reaction type, the use of imidazole-hydrazone as ligand was recently informed; the reaction was developed in short reaction times, improving the yields, all this result afforded by the use of NIR as the activating mode of reaction, with water as the solvent, offering a new ecological alternative to perform this kind of reaction, Entry 6.

2.8 Natural product extraction

Currently, it is well known that several methods are available to extract the secondary metabolites present in vegetable species, but recently, it was reported that the FIR can be also used [1] (**Table 8**).

E	Vegetable specimen	Extracted compounds	Ref.
1	*Oryza sativa* L.	*p*-Coumaric acid, 3-vinyl-1-oxybenzene, 4-hydroxybenzaldehyde, among others	[55, 56]
2	*Sesamum indicum* L.	*p*-Hydroxybenzoic acid, *o*-coumaric acid, vanillic acid, *p*-coumaric acid, isofeluric acid, sesamol, and tocopherol	[57]
3	*Arachis hypogaea* L.	2-Methoxyphenol, 2-hydroxy-4-methoxybenzoic acid, among others	[58]
4	*Camellia sinensis* var. sinensis	Polyphenolic compounds, caffeine, (−)-epicatechin, among others	[59–61]
5	*Puerariae radix*	Polyphenols compounds, isoflavonoids compounds	[62]
6	*Morus alba* L.	*p*-Coumaric acid, benzoic acid, (+)-catechin, chlorogenic acid, among others	[63]
7	*Radix Salviae miltiorrhizae*	Danshensu, protocatechuic acid, protocatechuic aldehyde, among others	[64]
8	*Lycium barbarum* Linn	Quercetin, rutin, and gentisic acid	[65]
9	*Glycyrrhiza uralensis* Fisch	Liquiritin, glycyrrhetic acid, and glycyrrhizin	[66]
10	*Lycium chinensis* Mill	Mannitol, sucrose, glucose, and fructose	[67]
11	Grape seeds	Catechin, epicatechin, and procyanidin B2	[68]
12	*Rhododendron mucronulatum* Turcz	Rutin, farrerol, syringic acid, vanillic acid, and 4-hydroxybenzoic acid	[69]

E	Vegetable specimen	Extracted compounds	Ref.
13	*Tagetes erecta* L.	*p*-Coumaric acid, gallic acid, among others	[70]
14	*Picrorhiza scrophulariiflora* Pennell	Picroside I and picroside II	[71]
15	*Hibiscus cannabinus* L.	Polyphenolic and flavonoids compounds, kaempferitrin, among others	[72]
16	*Glycine max* M.	Daidzein, and genistein	[73]
17	*Fagopyrum* spp.	Polyphenols and flavonoids	[74]
18	Apricot	Polyphenolic, flavonoids, and tannins compounds	[75]
19	Habanero pepper	Capsaicin and dihydrocapsaicin	[76]

E = Entry, Ref = Reference.

Table 8.
Extraction of natural compounds.

Interesting and recent information corresponds to a study by Cheaib and coworkers, Entry 18, reporting the extraction comparing the solid/liquid extraction with infrared-assisted extraction. In this study, the FIR pomace extract, from apricot, gave the highest polyphenolic content, flavonoid, and tannin yields. In addition, the inhibitory activity study against gram-positive and gram-negative bacteria was performed, and the infrared pomace gave the highest activity in comparison with infrared kernel, solid-liquid pomace, and solid-liquid kernel.

More recently, Entry 19, Martínez and coworkers have reported the extraction of capsaicin and dihydrocapsaicin, from habanero pepper in a comparative study, the use of NIR as source of extraction with EtOH as solvent, showed the best yields, 43.88 and 29.44% for capsaicin, and dihydrocapsaicin, respectively. The effect of nonconventional energies was analyzed by means of SEM micrographs, resulting in an NIR-procedure with a higher number of intact particles in the cellular matrix, in comparison with nonirradiated material.

3. Unpublished results from our and other laboratories

The results showed in this section correspond to novel results, being important to note that they have been recently submitted for publication or are under writing.

A green contribution to produce SiO_2–TiO_2 catalyst is offered, using NIR in comparison with other nonconventional energy sources and mantle heating. According to the obtained results, the materials were produced in short time reaction employing water as solvent. The results, for NIR strategy, show that smaller crystal size (15.3 nm) and higher BET surface area were obtained (335.2 m^2/g). In addition, the surface area obtained by mean NIR procedure, in comparison with other process, is due to that the water in the reaction was partially evaporated, affecting the hydrolysis process, but the pore and volume diameter was improved (4.33 nm and 0.363 cm^3/g, respectively) in comparison with the MH mode(2.77 nm and 0.290 cm^3/g) [77].

Related to natural product extraction, Miranda et al. have achieved a wide study about the extraction of perezone, from roots of both *Acourtia adnata* and *Acourtia platyphylla* comparing various modes: MH, NIR, MW, US, and supercritical CO_2. It is appropriated to comment that the yield obtained from the extraction with

NIR for 15 minutes was statically equal to the conventional thermal extraction, reflux during 3 h [78, 79].

Penieres-Carrillo and coworkers currently have developed the acylation of aromatic amines and alcohols, by a comparative study among MH, US, IR, and IR-US, with preliminary results; the best process has been the simultaneous use of IR-US, with high yields and shorter reaction times [80].

4. Conclusions

The request, "Decade to Educate in the Sustainability," established by the UNESCO (2005–2014) satisfactorily evaluated in November 2014, at Aichi, Nagoya, Japan is attended in this chapter, in this sense by means of a green approach using the infrared irradiation as an alternative and efficient mode to activate an organic reaction in addition to conveniently performed extractions of secondary metabolites. Moreover, it is important to note that this green energy can be used in other areas as material engineering and other synthetic strategies as Suzuki-Miyaura or Mizoroki-Heck cross-coupling can use it to activate the reaction. Consequently, the infrared irradiation can extend their application to other organic transformations or disciplines.

Acknowledgements

We would like to thank the Laboratorio de Estudios sobre Quimica Verde, L-122, Departamento de Ciencias Quimicas, Facultad de Estudios Superiores Cuautitlan, Universidad Nacional Autonoma de Mexico for financial affordings. R. Miranda Acknowledges PAPIME-UNAM PE 209919 for financial support.

Conflict of interest

The authors declare no conflict of interest.

Author details

Joel Martínez[1]* and René Miranda[2]

1 Chemistry Science Faculty, Autonomous University of San Luis Potosi, San Luis Potosi, Mexico

2 Department of Chemistry, Faculty of Superior Studies Cuautitlan, Campus 1, Autonomous National University of Mexico, Cuautitlan Izcalli, State of Mexico, Mexico

*Address all correspondence to: atlanta126@gmail.com

References

[1] Escobedo R, Miranda R, Martínez J. Infrared irradiation: Toward green chemistry, a review. International Journal of Molecular Sciences. 2016;**17**: 453. DOI: 10.3390/ijms17040453 and references therein.

[2] Learning to Live Together Sustainably (SDG4.7): Trends and Progress. Available from: http://en.unesco.org [Accessed: November 15, 2018]

[3] Education for Sustainable Development. Available from: http://en.unesco.org [Accessed: November 15, 2018]

[4] Anastas PT, Warner JC. Green Chemistry: Theory and Practice. 1st ed. New York: Oxford University Press; 1998. 135 p

[5] Lancaster M. Green Chemistry an Introductory Text. 1st ed. Cambridge: Royal Society Chemistry; 2007. pp. 5-19

[6] Doble M, Kruthiventi AK. Green Chemistry and Engineering. 1st ed. Burlington: Academic Press; 2007. pp. 27-37

[7] Erythropel HC, Zimmerman JB, de Winter TM, Petitjean L, Melnikov F, Lam CH, et al. The Green chemisTree: 20 years after taking root with the 12 principles. Green Chemistry. 2018;**20**: 1929-1961. DOI: 10.1039/C8GC00482J

[8] Shortwave Electric Infrared the Facts. Copyrigth© ITW BGK Finishing Systems 05/03. Available from: http://www.bgk.com/ir-the-facts [Accessed: November 15, 2018]

[9] Miranda R, Ríos H, Delgado M, Cogordán JA, Salmón J. Characterization of TAFF, a bentonitic clay and its application in the obtention of oligotoluene. Journal of Applied Catalyst. 2003;**224**:217-233

[10] Delgado F, Tamariz J, Zepeda G, Landa M, Miranda R, García J. Knoevenagel condensation catalyzed by a Mexican bentonite using infrared irradiation. Synthetic Communications. 1995;**25**:753-759

[11] Obrador E, Castro M, Tamaríz J, Zepeda G, Miranda R, Delgado F. Knoevenagel condensation in heterogeneous phase catalyzed by IR radiation and tonsil actisil FF. Synthetic Communications. 1998;**28**:4649-4663. DOI: 10.1080/00397919808004530

[12] Alcerreca G, Sanabria R, Miranda R, Arroyo G, Tamariz J, Delgado F. Preparation of benzylidene barbituric acids promoted by infrared irradiation in absence of solvent. Synthetic Communications. 2000;**30**:1295-1301. DOI: 10.1080/00397910008087151

[13] Penieres G, Miranda R, García J, Aceves J, Delgado F. Modification of the Fischer indole synthesis. Heterocyclic Communications. 1996;**2**:401-402

[14] Penieres G, Soto V, Alvarez C, García O, García JG. A new strategy for the synthesis of *N*-substituted 2,5-dimethylpyrroles in heterogeneous medium. Heterocyclic Communications. 1998;**4**:31-32

[15] Zhang C, Wang J, Li JH. Infrared heat aided solid state synthesis of pyrroles from 1,4-diketones and ammonium acetate. Journal of Heterocyclic Chemistry. 2012;**49**: 204-207. DOI: 10.1002/jhet.744

[16] Penieres CG, Bonifas AI, López CJG, García EJG, Alvarez TC. Synthesis of benzimidazoles in dry medium. Synthetic Communications. 2000;**30**: 2191-2195. DOI: 10.1080/00397910008087397

[17] Dandia A, Parewa V, Gupta SL, Rathore KS. Cobalt doped ZnS

nanoparticles as a recyclable catalyst for solvent-free synthesis of heterocyclic privileged medicinal scaffolds under infrared irradiation. Journal of Molecular Catalysis A: Chemical. 2013; **373**:61-71. DOI: 10.1016/j.molcata.2013.02.010

[18] Vázquez MA, Landa M, Reyes L, Miranda R, Tamariz J, Delgado F. Infrared irradiation: Effective promoter in the formation of *N*-benzylideneanilines in the absence of solvent. Synthetic Communications. 2004;**34**:2705-2718. DOI: 10.1081/SCC-200026190

[19] Tong JY, Sun NB, Wu HK. Grinding synthesis of Schiff bases combined with infrared irradiation. Asian Journal of Chemistry. 2013;**25**:5399-5401. DOI: 10.14233/ajchem.2013.14382

[20] Martínez J, Velasco-Bejarano B, Delgado F, Pozas R, Torres Domínguez HM, Trujillo Ferrara JG, et al. Eco-contribution to the chemistry of perezone, a comparative study, using different modes of activation and solventless conditions. Natural Product Communications. 2008;**3**:1465-1468

[21] Martínez J, Sánchez L, Pérez FJ, Carranza V, Delgado F, Reyes L, et al. Uncatalysed production of coumarin-3-carboxylic acids: A green approach. Journal of Chemistry. 2016;**2016**:1-6. DOI: 10.1155/2016/4678107

[22] Escobedo-González RG, Pérez H, Nicolás-Vázquez MI, Martínez J, Gómez G, Nava J, et al. Green production of indolylquinones, derivatives of perezone and related molecules, promising antineoplastic compounds. Journal of Chemistry. 2016;**2016**:1-10. DOI: 10.1155/2016/3870529

[23] Miranda R, Valencia-Vázquez O, Maya-Vega CA, Nicolás-Vázquez I, Vargas-Rodriguez YM, Morales-Serna JA, et al. Synthesis of cycloveratrylene macrocycles and benzyl oligomers catalysed by bentonitic under microwave/infrared and solvent-free conditions. Molecules. 2013;**18**: 12820-12844

[24] Salmón M, Osnaya R, Gómez L, Arroyo G, Delgado F, Miranda R. Contribution to the Biginelli reaction, using a bentonitic clay as catalyst and solventless procedure. Journal of the Mexican Chemical Society. 2001;**45**: 206-207

[25] Osnaya R, Arroyo GA, Parada L, Delgado F, Trujillo J, Salmón M, et al. Biginelli *vs* Hantzsch esters study under infrared radiation and solventless conditions. ARKIVOC. 2003;**2003**: 112-117

[26] Gómez-Pliego R, Osnaya R, Zamora I, Velasco-Bejarano B, Arroyo G, Ramírez-San Juan E, et al. The Hantzsch ester production in a water-based biphasic medium, using infrared irradiation as the activating source. Journal of the Mexican Chemical Society. 2007;**51**:181-184

[27] Martínez J, Romero-Vega S, Abeja-Cruz R, Álvarez-Toledano C, Miranda R. Green approach multicomponent production of boron containing Hantzsch and Biginelli esters. International Journal of Molecular Sciences. 2013;**14**:2903-2915. DOI: 10.3390/ijms14022903

[28] Chandrachood P, Gadkari T, Deshpande N, Kashalkar R. Highly efficient methodology for aromatization of 1,4-dihydropyridines using silica-supported transition metal nitrates under infra-red radiations. Journal of the Iranian Chemical Society. 2012;**9**:47-51. DOI: 10.1007/s13738-011-0010-2

[29] Noguez MO, Marcelino V, Rodríguez H, Martín O, Martínez JO, Arroyo GA, et al. Infrared assisted

production of 3,4-dihydro-2(1*H*)-pyridones in solvent-free conditions. International Journal of Molecular Sciences. 2011;**9**:47-51. DOI: 10.3390/ijms12042641

[30] Penieres-Carrillo G, García-Estrada JG, Gutiérrez-Ramírez JL, Alvarez-Toledano C. Infrared-assisted eco-friendly selective synthesis of diindolylmethanes. Green Chemistry. 2003;**5**:337-339. DOI: 10.1039/b211011c

[31] Wang SX, Guo SB, Gao MZ, Li JT, Duan YF. Infrared irradiation synthesis of substituted 5-oxo-1,2,3,4,5,6,7,8-octahydroquinoline derivatives under solvent-free conditions. E-Journal of Chemistry. 2006;**3**:159-163

[32] Sánchez A, Hernández F, Cruz PC, Alcaraz Y, Tamariz J, Delgado F, et al. Infrared irradiation- assisted multicomponent synthesis of 2-amino-3-cyano-4H-pyran derivatives. Journal of the Mexican Chemical Society. 2012; **56**:121-127

[33] Flores-Conde MI, Reyes L, Herrera R, Rios H, Vazquez MA, Miranda R, et al. Highly region- and stereoselective Diels-Alder cycloadditions via two-step and multicomponent reactions promoted by infrared irradiation under solvent-free conditions. International Journal of Molecular Sciences. 2012;**13**: 2590-2617. DOI: 10.3390/ijms13032590

[34] Valdez-Rojas JE, Ríos-Guerra H, Ramírez-Sánchez AL, García-González G, Álvarez-Toledano C, López-Cortés JG, et al. A study of the Willgerodt-Kindler reaction to obtain thioamides and α-ketothioamides under solvent-less conditions. Canadian Journal of Chemistry. 2012;**90**:567-573. DOI: 10.1139/V2012-030

[35] Velázquez AM, Torres LA, Díaz G, Ramírez A, Hernández R, Santillán H, et al. A novel one-pot, solvent-free Mannich synthesis of methylpiperidinyl phenols, methylphenylmorpholinyl phenols and methylthiophenyl morpholinyl phenols using infrared light irradiation. ARKIVOC. 2006;**2006**: 150-161

[36] Velázquez AM, Torres LA, González R, Valencia A, Díaz-Barriga S, Menconi I, et al. Synthesis of 4-methoxy-2-thiomorpholin-4-ylmethyl-1-phenol. Molbank. 2007;**2007**:M547

[37] Ortega-Jiménez F, Penieres-Carrillo JG, Lagunas–Rivera S, López-Cortés JG, Álvarez-Toledano C, Ortega-Alfaro MC. Infrared irradiation assisted both the synthesis of (Z)-(aminomethyl)(aryl) phenylhydrazones *via* the Mannich coupling reaction and their application to the palladium-catalyzed heck reaction. RSC Advances. 2015;**5**: 80911-80918. DOI: 10.1039/c5ra12715g

[38] Guo H, Peng M, Zhu Z, Sun L. Preparation of reduced grapheme oxide by infrared irradiation induced photothermal reduction. Nanoscale. 2013;**5**:9040-9048. DOI: 10.1039/c3nr02805d

[39] Wang G, Huang B, Ma X, Wang Z, Qin X, Zhang X, et al. $Cu_2(OH)PO_4$, a near-infrared-activated photocatalyst. Angewandte Chemie International Edition. 2013;**52**:4810-4813. DOI: 10.1002/anie.201301306

[40] Ikeue T, Sonoda M, Kurahashi S, Tachibana H, Teraoka D, Sugimori T, et al. Annulated dinuclearpalladium (II) phthalocyanine complex as an effective photo-oxidation catalyst for near-infrared region light. Inorganic Chemistry Communications. 2010;**13**:1170-1172. DOI: 10.1016/j.inoche.2010.06.041

[41] Gómez-Lara J, Gutierrez-Pérez R, Penieres-Carrillo G, López-Cortés JG, Escudero-Salas A, Álvarez-Toledano C. Reaction of hydroquinones with supported oxidizing reagents in solvent-free conditions. Synthetic

Communications. 2000;**30**:2713-2720. DOI: 10.1080/00397910008086896

[42] Li H, Liu R, Lian S, Liu Y, Huang H, Kang Z. Near-infrared light controlled photocatalytic activity of carbon quantum dots for highly selective oxidation reaction. Nanoscale. 2013;**5**: 3289-3297. DOI: 10.1039/c3nr00092c

[43] Deng CH, Li YM, Zhao YF. Formation of oligopeptides from *N*-phosphoamino acid by infrared radiation. Phosphorus, Sulfur, and Silicon and The Related Elements. 2000; **163**:203-210. DOI: 10.1080/10426500008046620

[44] Hernández MT, González M. Synthesis of resins as α-alumina precursors by the Pechini method using microwave and infrared heating. Journal of the European Ceramic Society. 2002; **22**:2861-2868

[45] Huang L, Wei HB, Ke FS, Cai JS, Fan XY, Sun SG. Infrared irradiation-assisted one-step synthesis of nanosized tin dioxide particles and particle size effect on lithium storage performance. Colloids and Surfaces A: Physicochemical and Engineering Aspects. 2007;**308**:87-92. DOI: 10.1016/j.colsurfa.2007.05.034

[46] Termoss H, Toury B, Pavan S, Brioude A, Bernard S, Cornu D, et al. Preparation of boron nitride-based coatings on metallic substrates via infrared irradiation of dip-coated polyborazylene. Journal of Materials Chemistry. 2009;**19**:2671-2674. DOI: 10.1039/b820327j

[47] Zhang N, Li Z, Zheng J, Yang X, Shen K, Zhou T, et al. Multielemental analysis of botanical samples by ICP-OES and ICP-MS with focused infrared lightwave ashing for sample preparation. Microchemical Journal. 2017;**134**:68-77. DOI: 10.1016/j.microc.2017.05.006

[48] Jiménez-Estrada M, García MO, Navarro AO, Eusebio JL, Álvarez CT, Penieres GC, et al. Allylic nitration of 3-sitosterol and cholesterol acetate: Preparation of 7-nitro derivatives. Steroids. 1997;**62**:500-503

[49] Penieres G, Aceves JM, Flores A, Mendoza G, García O, Álvarez C. Comparative study of ε-caprolactam synthesis using different energy sources and natural clay as catalyst. Heterocyclic Communications. 1997;**3**:507-508

[50] Luna-Mora RA, Torres-Reyes A, González-Cruz OA, Ortega-Jiménez F, Ríos-Guerra H, González-Carrillo JV, et al. Assessment of amination reactions via nucleophilic aromatic substitution using conventional and eco-friendly energies. Green Chemistry Letters and Reviews. 2018;**11**:371-378

[51] Ortega-Jiménez F, Domínguez-Villa FX, Rosas-Sánchez A, Penieres-Carrillo G, López-Cortés JG, Ortega-Alfaro MC. An expedient approach to enhance Mizoroki-Heck coupling reaction by infrared irradiation using palladacycle compounds. Applied Organometallic Chemistry. 2015;**29**:556-560. DOI: 10.1002/aoc.3331

[52] Balam-Villareal JA, Sandoval-Chávez CI, Ortega-Jiménez F, Toscano RA, Carréon-Castro MP, López-Cortés JG, et al. Infrared irradiation or microwave assisted cross-coupling reactions using sulfur-containing ferrocenyl-palladacycles. Journal of Organometallic Chemistry. 2016;**818**: 7-14. DOI: 10.1016/j.jorganchem.2016.05.017

[53] Ortega-Jiménez F, Penieres-Carrillo JG, López-Cortés JG, Ortega-Alfaro MC, Lagunas-Rivera S. Arylhydrazones derivatives containing a benzothiazole moiety, efficient ligands in the palladium-catalyzed Mizoroki-Heck and Suzuki-Miyaura cross-coupling reactions under IR irradiation. Chinese

Journal of Chemistry. 2017;**35**: 1881-1888. DOI: 10.1002/cjoc.201700390

[54] Camacho-Espinoza M, Penieres-Carrillo JG, Rios Guerra H, Lagunas-Rivera S, Ortega-Jiménez F. An efficient and simple imidazole-hydrazone ligand for palladium-catalyzed Suzuki-Miyaura cross coupling reactions in water under infrared irradiation. Journal of Organometallic Chemistry. 2019;**880**: 386-391. DOI: 10.1016/j.jorganchem.2018.11.016

[55] Lee SC, Kim JH, Jeong SM, Kim DR, Ha JU, Nam KC, et al. Effect of far-infrared radiation on the antioxidant activity of rice hulls. Journal of Agricultural and Food Chemistry. 2003;**51**:4400-4403. DOI: 10.1021/jf0300285

[56] Lee SC, Kim JH, Nam KC, Ahn DU. Antioxidant properties of the far infrared-treated rice hull extract in irradiated raw and cooked turkey breast. Journal of Food Science. 2003;**68**: 1904-1909

[57] Lee SC, Jeong SM, Kim SY, Nam KC, Ahn DU. Effect of far-infrared irradiation on the antioxidant activity of defatted sesame meal extracts. Journal of Agricultural and Food Chemistry. 2005;**53**:1495-1498. DOI: 10.1021/jf048620x

[58] Lee SC, Jeong SM, Kim SY, Park HR, Nam KC, Ahn DU. Effect of far-infrared radiation and heat treatment on the antioxidant activity of water extract from peanut hulls. Food Chemistry. 2006;**94**:489-493. DOI: 10.1016/j.foodchem.2004.12.001

[59] Lee SC, Kim SY, Jeong SM, Park JH. Effect of far-infrared irradiation on catechins and nitrile scavenging activity of green tea. Journal of Agricultural and Food Chemistry. 2006;**54**:399-403. DOI: 10.1021/jf051866x

[60] Kim SY, Jeong SM, Jo SC, Lee SC. Application of far-infrared irradiation in the manufacturing process of green tea. Journal of Agricultural and Food Chemistry. 2006;**54**:9943-9947. DOI: 10.1021/jf062524+

[61] Park JH, Lee JM, Cho YJ, Kim CT, Kim CJ, Nam KC, et al. Effect of far-infrared heater on the physicochemical characteristics of green tea during processing. Journal of Food Biochemistry. 2009;**33**:149-162

[62] Kim JW, Bae HC, Kim MC, Lee SC. Effect of far-infrared irradiation on the antioxidant activity of *Puerariae radix* extract. Journal of Food Biochemistry. 2008;**32**:85-95

[63] Wanyo P, Siriamornpun S, Meeso N. Changes in phenolic compounds, antioxidant and physical properties of mulberry tea influenced by intensity of far-infrared radiation. American-Eurasian Journal of Agricultural & Environmental Sciences. 2009;**6**: 470-479

[64] Chen Y, Duan G, Xie M, Chen B, Li Y. Infrared-assisted extraction coupled with high-performance liquid chromatography for simultaneous determination of eight active compounds in *Radix salviae* miltiorrhizae. Journal of Separation Science. 2010;**33**:2888-2897. DOI: 10.1002/jssc.201100649

[65] Duan H, Chen Y, Chen G. Far infrared-assisted extraction followed by capillary electrophoresis for the determination of bioactive constituents in the leaves of *Lycium barbarum* Linn. Journal of Chromatography. A. 2010; **1217**:4511-4516. DOI: 10.1016/j.chroma.2010.04.069

[66] Lee JM, Lee SC. The effects of far-infrared irradiation on the antioxidant activity of licorice (*Glycyrrhiza uralensis*

Fisch) root. Journal of Food Biochemistry. 2010;**34**:172-181. DOI: 10.1111/j.1745-4514.2009.00272.x

[67] Fu Y, Zhang L, Chen G. Determination of carbohydrates in Folium *Lysium chinensis* using capillary electrophoresis combined with far-infrared light irradiation-assisted extraction. Journal of Separation Science. 2011;**34**:3272-3278. DOI: 10.1002/jssc.201100649

[68] Cai Y, Yu Y, Duan G, Li Y. Study on infrared-assisted extraction coupled with high performance liquid chromatography (HPLC) for determination of catechin, epicatechin, and procyanidin B2 in grape seeds. Food Chemistry. 2011;**127**:1872-1877. DOI: 10.1016/j.foodchem.2011.02.026

[69] Fu Y, Zhang L, Chen G. Far infrared-assisted extraction followed by MEKC for the simultaneous determination of flavones and phenolic acids in the leaves of *Rhododendron mucronulatum* Turcz. Journal of Separation Science. 2012;**35**:468-475. DOI: 10.1002/jssc.201100816

[70] Siriamornpun S, Kaisoon O, Meeso N. Changes in colour, antioxidant activities and carotenoids (lycopene, β-carotene, lutein) of marigold flower (*Tagetes erecta* L.) resulting from different drying processes. Journal of Functional Foods. 2012;**4**:757-766. DOI: 10.1016/j.jff.2012.05.002

[71] Li F, Yu Y, Zhang H, Liu T, Li Y, Duan G. Infrared-assisted non-ionic surfactant extraction as a green analytical preparatory technique for the rapid extraction and pre-concentration of pricoside I and picroside II from *Picrorhiza scrophulariiflora* Pennell. Analytical Methods. 2013;**5**:3747-3753. DOI: 10.1039/c3ay40135a

[72] Jin CW, Ghimeray AK, Wang L, Xu ML, Piao JP, Cho DH. Far infrared assisted Kenaf leaf tea preparation and its effect on phenolic compounds, antioxidant and ACE inhibitory activity. Journal of Medicinal Plant Research. 2013; 7:1121-1128. DOI: 10.5897/JMPR11.1431

[73] Ghimeray AK, Sharma P, Hu W, Jin CW, Park CH, Rho HS, et al. Far infrared assisted conversion of isoflavones and its effect on total phenolics and antioxidants activity in black soybean seed. Journal of Medicinal Plant Research. 2013;7:1129-1137. DOI: 10.5897/JMPR11.1507

[74] Ghimeray AK, Sharma P, Phoutaxay P, Salitxay T, Woo SH, Park CH. Far infrared irradiation alters total polyphenol, total flavonoid, antioxidant property and quercetin production in tartary buckwheat sprout powder. Journal of Cereal Science. 2014;**59**: 167-172. DOI: 10.1016/j.jcs.2013.12.007

[75] Cheaib D, Darra NE, Rajha HN, Ghazzawi IE, Maroun RG, Louka N. Biological activity of apricot byproducts polyphenols using solid-liquid and infrared-assisted technology. Journal of Food Biochemistry. 2018;**42**:e12552. DOI: 10.1111/jfbc.12552

[76] Martínez J, Rosas J, Pérez J, Saavedra Z, Carranza V, Alonso P. Green approach to the extraction capsaicinoids from habanero pepper using near-infrared, microwave, ultrasound and Soxhlet methods, a comparative study. Natural Product Research. 2018. DOI: 10.1080/14786419.2018.1455038

[77] Alonso P, Pérez Salas H, Miranda R, Zarraga R, Moctezuma E, Saavedra Z, et al. Green approach for the sol-gel production of SiO_2-TiO_2 mesoporous catalysts, comparing various activating reaction modes. Green Chemistry. 2018. Submitted

[78] Miranda R et al. Green approach extraction of perezone from *Acourtia adnata*, unpublished results. Laboratorio de Estudios sobre Quimica Verde, L-122. Departamento de Ciencias Quimicas, Facultad de Estudios Superiores Cuautitlan, Universidad Nacional Autonoma de Mexico

[79] Miranda R et al. Green approach extraction of perezone from *Acourtia platyphilla*, unpublished results. Laboratorio de Estudios sobre Quimica Verde, L-122. Departamento de Ciencias Quimicas, Facultad de Estudios Superiores Cuautitlan, Universidad Nacional Autonoma de Mexico

[80] Luna–Mora RA, Ortega-Jiménez F, Ríos-Guerra H, García-Estrada JG, Pérez-Flores FJ, González-Carrillo J, et al. Simultaneous infrared-ultrasound irradiation in organic synthesis: Acylation of amines and alcohols. Journal of the Mexican Chemical Society. 2018. Submitted

Chapter 4

PEG-Mediated Green One Pot Synthesis by Using Click Chemistry

Sachin Pandurang Shirame and Raghunath Bhosale

Abstract

The regioselective synthesis of 1,4-disubstituted 1,2,3-triazoles derivatives from substituted alkynes and organic halides with sodium azide by using CuI catalyst in polyethylene glycol-400 as a green reaction media. This process is of considerable synthetic advantages in terms of green principles, high atom economy, low environmental impact, mild reaction condition, high purity and good yields. We find out the use of eco-friendly solvent like mixture of PEG-400 and water for the synthesis of 1,2,3-triazole. The main aim of this research is to found the method which required very short time, cost effective, feasible and a green method as compared to known reported for synthesis of 1,2,3-triazole as a medicinally important scaffold by click chemistry.

Keywords: PEG-400, multicomponent reactions, 1,2,3-triazole, CuI

1. Introduction

'Click chemistry' has emerged as a fast and efficient approach for synthesis of novel heterocyclic compounds [1, 2]. The Huisgen 1,3-dipolar cycloaddition of azides and alkynes resulting in 1,2,3-triazoles is one of the most powerful click reactions [3, 4]. The synthesis of 1,2,3-triazole has been intensively studied, and triazoles are widely used in pharmaceuticals, agrochemicals, dyes, photographic materials, and in corrosion inhibitory materials [5–7]. In addition, they possesses anti-HIV [8, 9] antimicrobial activities [10]. The selective β-3 adrenergic receptor agonism [11]. In the absence of a transition-metal catalyst, these reactions are not regioselective, relatively slow, and require high temperatures to reach acceptable yields. In early 2002, Meldal and co-workers reported that the use of catalytic amounts of copper(I), which can bind to terminal alkynes, leads to fast, highly efficient, and regioselective azide, alkyne cycloadditions at room temperature in organic medium [12–15]. Recently, Sharpless and co-workers have reported a high yielding synthesis of triazoles using a CuI catalyst with an excellent 1,4-regioselectivity [15–18]. The resulting 'clicked' products can even be obtained via in situ generation of the corresponding organic azides from organic halides, NaN_3 in the presence of an alkyne and a copper catalyst, avoiding the need to handle organic azides [19]. Nitrogen heterocycles have received special attention in pharmaceutical chemistry due to their diverse medicinal potential [20–22]. The main aim of our research work is to replace the costly and hazardous organic solvents for the synthesis of 1,2,3-triazoles by using ecofriendly efficient unique properties such as commercial

IntechOpen

availability, recyclable, easily degradable, having low toxicity, thermally stability and non-volatility of this PEG-400 solvent [23].

2. Materials and method

All chemicals were purchased from Merck and Aldrich and used as received. Melting points were recorded in open capillaries.^{1}H NMR were recorded on a Bruker Bio-Spin spectrometer at 400 MHz using TMS as an internal standard (in $CDCl_3$). Mass spectra ESIMS were recorded and IR spectra were recorded on a Shimadzu FTIR spectrometer in KBr pallets.

3. General procedure for the synthesis of 1,4-disubstituted 1,2,3-triazole for compounds (111)

Substituted organic halides (1.0 equiv), sodium azide (1.4 equiv) and substituted alkynes (1.104 equiv) were suspended in polyethylene glycol-400 (5 mL). To this copper iodide (10 mol%) was added and the re action mixture was stirred for 10–45 min at 25–35°C. After completion of the reaction (monitored by TLC), the product was extracted with EtOAc and the organic extract was dried. The crude product was recrystallized from ethanol to yield the desired product.

4. Experimental data

4.1 Synthesis of 1-(4-nitrobenzyl)-4-phenyl-1H-1,2,3-triazole

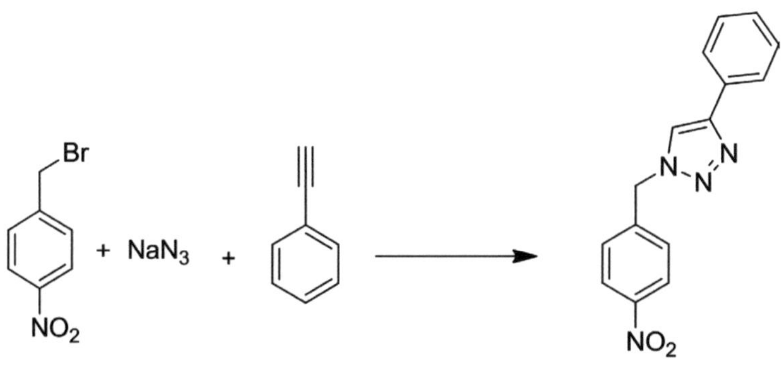

P-Nitrobenzyl bromide 0.216 g (1.0 mmol), sodium azide 0.091 g (1.4 mmol) and phenyl acetelyne 0.112 g (1.1 mmol) were suspended in polyethylene glycol-400 (5 mL). To this reaction mixture in copper iodide (10 mol%) was added and the reaction mixture was stirred for 10 min at 25°C. After completion of the reaction (monitored by TLC), the product was extracted with EtOAc and the organic extract was dried. The crude product was recrystallized from ethanol to yield the desired product.

MF/FWt: $C_{15}H_{12}O_2N_4$/280.10, MP: 198–200°C.

IR (cm^{-1}): 3001, 2988, 2829, 1613, 1209, 876.

^{1}H NMR (300 MHz, $CDCl_3$, δ ppm): 8.13–8.15 (d,2H), 7.08–7.10 (d,2H),7.80–7.82 (d,2H), 6.85–6.87 (d,2H), 7.10–7.12 (s,1H), 8.39 (s,1H, triazole), 4.73 (s,2H).

4.2 Synthesis of 1-(4-nitrobenzyl)-4-p-tolyl-1H-1,2,3-triazole

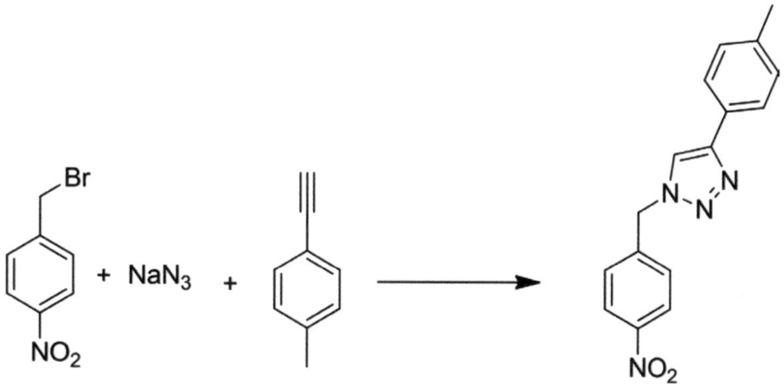

P-Nitrobenzyl bromide 0.216 g (1.0 mmol), sodium azide 0.091 g (1.4 mmol) and 1-ethynyl-4-methylbenzene 0.127 g (1.1 mmol) were suspended in polyethylene glycol-400 (5 mL). To this reaction mixture in copper iodide (10 mol%) was added and the reaction mixture was stirred for 15 min at 30°C. After completion of the reaction (monitored by TLC), the product was extracted with EtOAc and the organic extract was dried. The crude product was recrystallized from ethanol to yield the desired product.

Yield: 87%, MF/FWt: $C_{16}H_{14}N_4O_2$/294.11, MP: 190–192°C.

IR (cm^{-1}): 3123, 2958, 1588, 1430, 1265, 1233, 797.

^{1}H NMR (300 MHz, $CDCl_3$, δ ppm): 8.13–8.15 (d,2H), 7.08–7.10 (d,2H), 7.80–7.82 (d,2H), 6.85–6.87 (d,2H), 8.39 (s,1H, triazole), 4.73 (s,2H), 2.56 (s,3H).

4.3 Synthesis of (1-(4-nitrobenzyl)-1H-1,2,3-triazole-4-yl) methanol

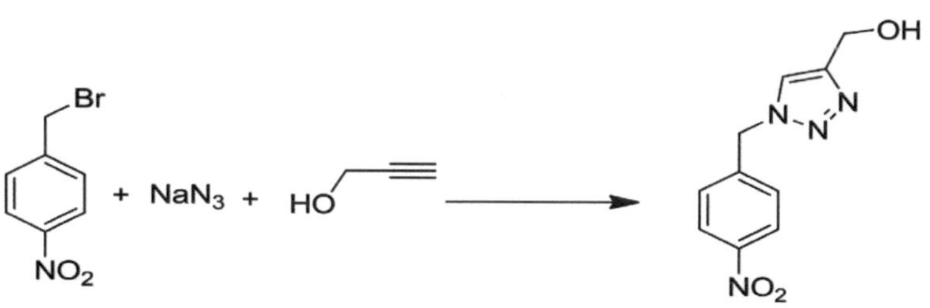

P-Nitrobenzyl bromide 0.216 g (1.0 mmol), sodium azide 0.091 g (1.4 mmol) was suspended in polyethylene glycol-400 (5 mL). The reaction mixture was stirred for 10 min at 25°C. After completion of the reaction (monitored by TLC), the product was extracted with EtOAc and the organic extract was dried. The crude product was recrystallized from ethanol to yield the desired product.

Yield: 72%, MF/FWt: $C_{10}H_{10}N_4O_3$/234.08, MP: 134–136°C.

IR (cm^{-1}): 3643, 2950, 1609, 1508, 1430, 1265, 1233, 867.

^{1}H NMR (300 MHz, $CDCl_3$, δ ppm): 8.13–8.15 (d,2H), 7.08–7.10 (d,2H), (s,1H, triazole), 4.73 (s,2H), 4.70 (s,2H), 3.65 (s,1H).

4.4 Synthesis of 1-benzyl-4-phenyl-1H-1,2,3-triazole

Benzyl bromide 0.171 g (1.0 mmol), sodium azide 0.091 g (1.4 mmol) and phenyl acetylene 0.112 g (1.1 mmol) were suspended in polyethylene glycol-400 (5 mL).

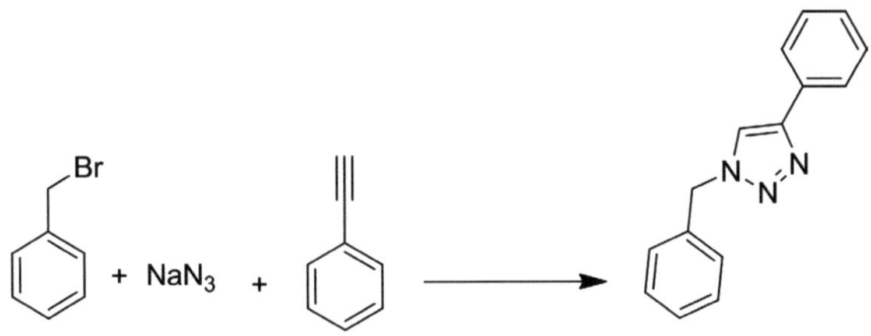

To this reaction mixture in copper iodide (10 mol%) was added and the reaction mixture was stirred for 15 min at 30°C. after completion of the reaction (monitored by TLC), the product was extracted with EtOAc and the organic extract was dried. The crude product was recrystallized from ethanol to yield the desired product.

Yield: 82%, MF/FWt: $C_{15}H_{13}N_3$/235.11, MP: 139–141°C.

IR (cm^{-1}): 2950, 1644, 1578, 1435, 1260, 1223, 876.

^{1}H NMR (300 MHz, $CDCl_3$, δ ppm): 6.87–6.89 (d,2H), 6.64–6.65 (t,1H), 7.12–7.13 (d,2H), 7.53–7.66 (d,2H), 7.39–7.42 (t,2H), 7.30–7.34 (t,1H), 7.66 (s,1H, triazole), 5.53 (s,2H).

4.5 Synthesis of 1,4-bromobenzyl, 4-phenyl-1H-1,2,3-triazole

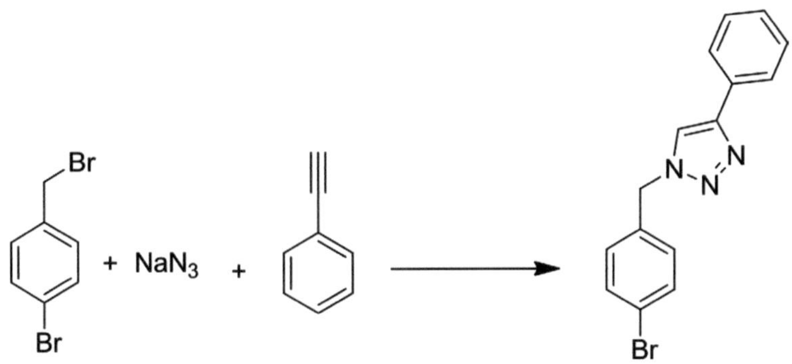

1-Bromo-4-(bromomethyl) benzene 0.249 g (1.0 mmol), sodium azide 0.091 g (1.4 mmol) and phenyl acetelyne 0.112 g (1.1 mmol) were suspended in polyethylene glycol-400 (5 mL). To this reaction mixture in copper iodide (10 mol%) was added and the reaction mixture was stirred for 20 min at 25°C. After completion of the reaction (monitored by TLC), the product was extracted with EtOAc and the organic extract was dried. The crude product was recrystallized from ethanol to yield the desired product.

Yield: 82%, MF/FWt: $C_{15}H_{12}BrN_3$/313.02, MP: 140–142°C.

IR (cm^{-1}): 2980, 1601, 1545, 1225, 1157.

^{1}H NMR (300 MHz, $CDCl_3$, δ ppm): 8.13–8.15 (d,2H), 7.08–7.10 (d,2H), 7.80–7.82 (d,2H), 6.85–6.87 (d,2H),7.81–7.83 (t,1H), 8.39 (s,1H, triazole), 4.73 (s,2H).

^{13}C NMR (300 MHz, $CDCl_3$, δ ppm): 53, 113, 114, 118, 133 (carbon triazole), 159, 148, 149, 141.

4.6 Synthesis of 1-4-bromobenzyl-1H-1,2,3-triazol-4-yl)methanol

1-Bromo-4-(bromomethyl) benzene 0.249 g (1.0 mmol), sodium azide 0.091 g (1.4 mmol) were suspended in polyethylene glycol-400 (5 mL). The reaction

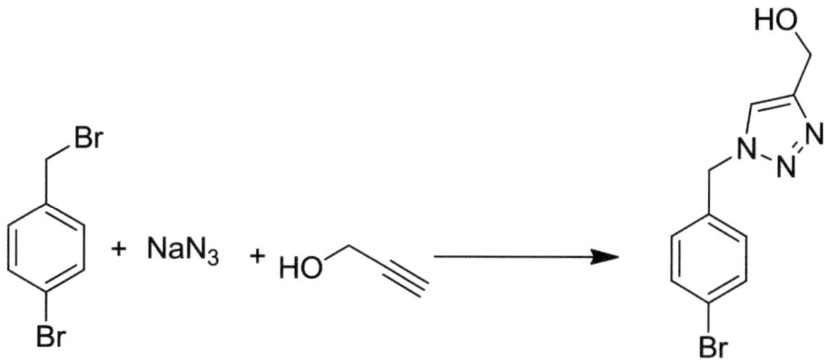

mixture was stirred for 1 h at 40–45°C. Then add propargyl alcohol 0.064 g (1.1 mmol), to this reaction mixture in copper iodide (10 mol%) was added and again reaction mixture was stirred for 10–45 min at 25–35°C. After completion of the reaction (monitored by TLC), the product was extracted with EtOAc and the organic extract was dried. The crude product was recrystallized from ethanol to yield the desired product.

Yield: 81%, MF/FWt: $C_{10}H_{10}BrN_3O$/267.00, MP: 148–150°C.

IR (cm^{-1}): 3660, 2950, 1643, 1578, 1435, 1262, 1222, 879.

^{1}H NMR (300 MHz, $CDCl_3$, δ ppm): 7.51–7.53(d,2H), 7.14–7.16 (d,2H), 7.73 (s,1H, triazole), 5.50 (s,2H), 4.48 (s,2H), 3.50 (s,1H).

4.7 Synthesis of 4-benzyloxy methyl-1-4-bromobenzyl-1H-1,2,3-triazole

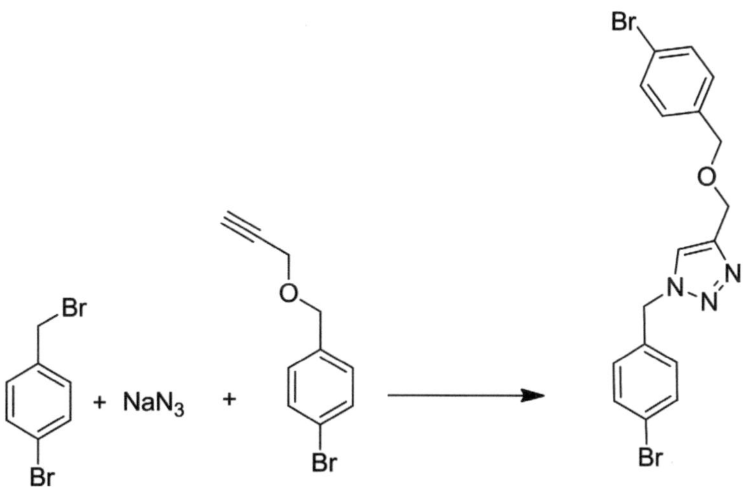

1-Bromo-4-(bromomethyl) benzene 0.249 g (1.0 mmol), sodium azide 0.091 g (1.4 mmol) and 1-bromo-4-((prop-2-ynyloxy)methyl)benzene 0.247 g (1.1 mmol) were suspended in polyethylene glycol-400 (5 mL). To this reaction mixture in copper iodide (10 mol%) was added and the reaction mixture was stirred for 10–45 min at 25–35°C. After completion of the reaction (monitored by TLC), the product was extracted with EtOAc and the organic extract was dried. The crude product was recrystallized from ethanol to yield the desired product.

Yield: 89%, MF/FWt: $C_{17}H_{16}BrN_3O$/357.05, MP: 160–162°C.

IR (cm^{-1}): 2945, 1640, 1588, 1435, 1262, 1222, 867.

^{1}H NMR (300 MHz, $CDCl_3$, δ ppm): 7.49–7.51 (d,2H), 7.19–7.21 (d,2H),7.30–7.32 (d,3H), 7.44–7.46 (d,2H),7.66 (s,1H, triazole), 5.47 (s,2H), 4.53 (s,2H), 4.64 (s,2H).

4.8 Synthesis of 1-(4-bromobenzyl)-4-(bromomethyl)-1H-1,2,3-triazole

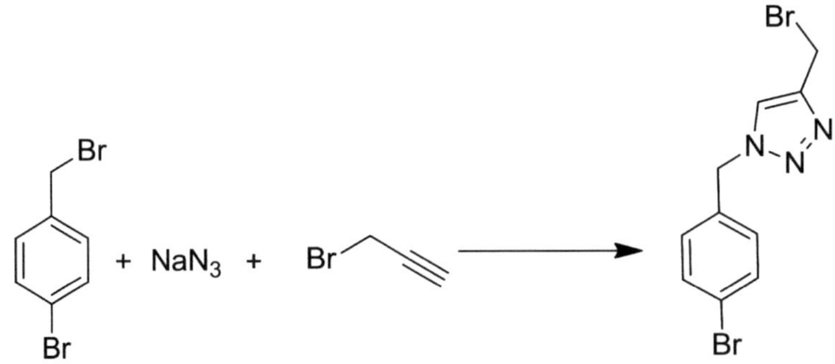

1-Bromo-4-(bromomethyl) benzene 0.249 g (1.0 mmol), sodium azide 0.091 g (1.4 mmol) were suspended in polyethylene glycol-400 (5 mL). The reaction mixture was stirred for 1 h at 40–45°C. Then add propargyl bromide 0.112 g (1.1 mmol), to this reaction mixture in copper iodide (10 mol%) was added and again reaction mixture was stirred for 20 min at 35°C. After completion of the reaction (monitored by TLC), the product was extracted with EtOAc and the organic extract was dried. The crude product was recrystallized from ethanol to yield the desired product.

Yield: 85%, MF/FWt: $C_{10}H_9Br_2N$/328.92, MP: 150–152°C.

IR (cm^{-1}): 2980, 1653, 1568, 1465, 1210, 1222, 889.

^{1}H NMR (300 MHz, $CDCl_3$, δ ppm): 7.14–7.16 (d,2H), 7.03–7.05 (d,2H),7.73 (s,1H, triazole), 5.50 (s,2H), 4.48 (s,2H).

4.9 Synthesis of 4-(4-pheny l-1H-1,2,3-treiazol-1yl) methyl)benzonitrile

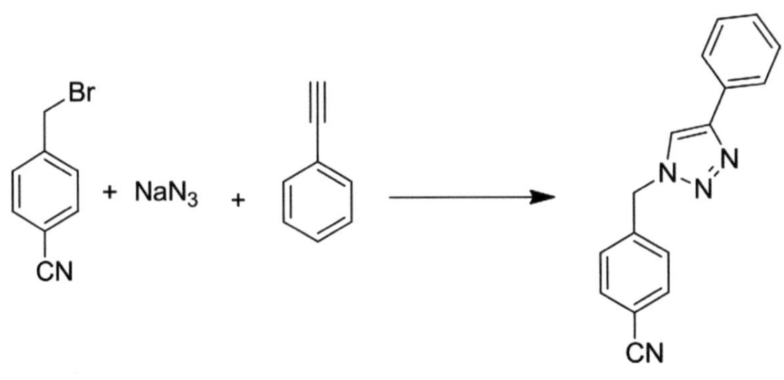

4-(Bromomethyl) benzonitrile 0.196 g (1.0 mmol), sodium azide 0.091 g (1.4 mmol) and phenyl acetelyne 0.112 g (1.1 mmol) were suspended in polyethylene glycol-400 (5 mL). To this reaction mixture in copper iodide (10 mol%) was added and the reaction mixture was stirred for 10 min at 25°C. After completion of the reaction (monitored by TLC), the product was extracted with EtOAc and the organic extract was dried. The crude product was recrystallized from ethanol to yield the desired product.

Yield: 7 8%, MF/FWt: $C_{16}H_{12}N_4$/260.11, MP: 130–132°C.

IR (cm^{-1}): 3088, 2921, 2850, 1607, 1488, 1026.

^{1}H NMR (300 MHz, $CDCl_3$, δ ppm): 7.66–7.68 (d,2H), 7.34–7.36 (t,2H), 7.38–7.43 (t,2H), 7.79–7.81 (d,2H), 7,26–7.32 (t,1H), 7.73 (s,1H, triazole), 5.64 (s,2H); MS: m/e 260 (M^{+1}).

4.10 Synthesis of 1-(4-isocyanobenzy l)-4-p-tolyl-1H-1,2,3-triazole

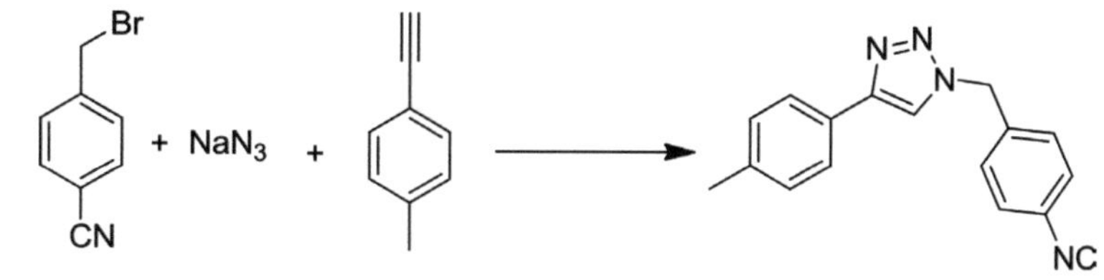

4-(Bromomethyl) benzonitrile 0.196 g (1.0 mmol), sodium azide 0.091 g (1.4 mmol) and 1-ethynyl-4-methylbenzene 0.127 g (1.1 mmol) were suspended in polyethylene glycol-400 (5 mL). To this reaction mixture in copper iodide (10 mol%) was added and the reaction mixture was stirred for 15 min at 30°C. After completion of the reaction (monitored by TLC), the product was extracted with EtOAc and the organic extract was dried. The crude product was recrystallized from ethanol to yield the desired product.

Yield: 87%, MF/FWt: $C_{17}H_{14}N_4$/274.12, MP: 133–135°C.

IR (cm^{-1}): 3081, 2223, 1669, 1470, 1219, 834.

^{1}H NMR (300 MHz, $CDCl_3$, δ ppm): 7.66–7.68(d,2H), 7.34–7.36 (d,2H), 7.79–7.81 (d,2H), 7.38–7.43 (d,2H), 7.73 (s,1H, triazole), 5.64 (s,2H), 2.58 (s,3H).

4.11 Synthesis of 1-(4-isocyanobenzyl)-4-p-tolyl-1H-1,2,3-triazole

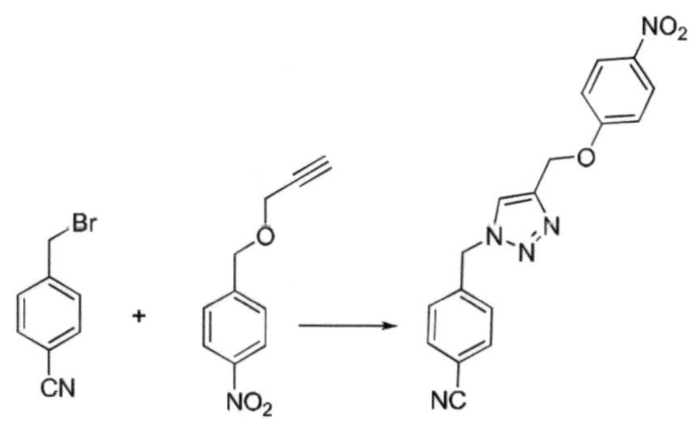

4-(Bromomethyl) benzonitrile 0.196 g (1.0 mmol), sodium azide 0.091 g (1.4 mmol) and 1-methyl-4-(prop-2-ynyloxy) benzene 0.160 g (1.1 mmol) were suspended in polyethylene glycol-400 (5 mL). To this reaction mixture in copper iodide (10 mol%) was added and the reaction mixture was stirred for 40 min at 35°C. After completion of the reaction (monitored by TLC), the product was extracted with EtOAc and the organic extract was dried. The crude product was recrystallized from ethanol to yield the desired product.

Yield: 7 8%, MF/FWt: $C_{18}H_{16}N_4O$/304.13, MP: 144–146°C.

IR (cm^{-1}): 3145, 2227, 1604, 1437, 1154, 857.

^{1}H NMR (300 MHz, $CDCl_3$, δ ppm): 7.66–7.68 (d,2H), 7.34–7.36 (d,2H), 7.43–7.45 (d,2H), 7.84–7.86 (d,2H), 7.73 (s,1H, triazole), 5.61 (s,2H), 5.37 (s,2H), 2.56 (s,3H).

5. Conclusions

In conclusion a safe and efficient method for the generation of 1,4-disubstituted 1,2,3-triazole in a complete regioselective manner has been developed. Synthesis of

1,2,3-triazole moiety is carried out for the first time by using PEG-400 as a green solvent. This methods are versatile, efficient and convenient. The methods required very short time as compared to reported methods for the synthesis of multicomponent 1,2,3-triazole and their heterocyclic compounds. Avoids the use of expensive volatile organic solvents and laborious work-up. Multicomponent method 1,2,3-triazole derivatives were synthesized. This method avoids isolation and handling of potentially unstable organic azide and provides triazole product in pure form 1,2,3-triazole moiety as a medicinal use.

Acknowledgements

The authors thank to the Prof. Dr. R.B. Bhosale, Director and Head, Department of Organic Chemistry, Solapur University and Prof. Miss. Fandnewis, Vice Chancellor, Solapur University for providing necessary laboratory facilities. SPS also moral support CEO Pramod M. Kawale from Goga Industry, Dhule.

List of abbreviations

CuI	copper iodide
NaN_3	sodium azide
EA	ethylacetate
CuAAC	copper-catalyzed azide-alkyne cycloaddition
DCC	*N,N′*-dicyclohexylcarbodiimide
FT-IR	Fourier transformation infra-red
NMR	nuclear magnetic resonance
PEG	poly(ethylene glycol)
RuAAC	ruthenium-catalyzed alky ne azide cycloaddition
THF	tetrahydrofuran
TLC	thin layer chromatography
MCR	multicomponent reaction synthesis
$NaNO_2$	sodium nitrite

Author details

Sachin Pandurang Shirame* and Raghunath Bhosale
School of Chemical Sciences, Solapur University, Solapur, India

*Address all correspondence to: sachinshirame@gmail.com

References

[1] Kolb HC, Finn MG, Sharpless KB. Angewandte Chemie International Edition. 2001;**40**

[2] Kolb HC, KB S. Drug Discovery Today. 2003;**8**:1128

[3] Huisgen R, Padwa A, editors. New York: Wiley; 1984. pp. 1

[4] Ming X, Leonard P, Heindl D, Seela F. Nucleic Acids Symposium Series. 2008;**52**:471

[5] Meldal M, Tornøe CW. Chemical Reviews. 2008;**108**:2952

[6] Nandivada H, Jiang X, Lahann J. Advanced Materials. 2007;**19**:2197

[7] Fan WQ, Katrisky AR. In: Katrisky AR, Rees CW, Scriven CW, editors. Comprehensive Heterocyclic Chemistry II. Vol. 4. Oxford: Elsevier Science; 1996. p. 1

[8] Alvarez S, San F, Aquaro S, Perno CF, Karlsson A, Balzarini J, et al. Journal of Medicinal Chemistry. 1994;**37**:4185

[9] Velazquez S, Alvarez R, Perez C, Gago F, De C, Balzarini J, et al. Journal Antiviral Chemistry and Chemotherapy. 1998;**9**:481

[10] Genin MJ, Allwine DA, Anderson DJ, Barbachyn MR, Emmert DE, Garmon SA, et al. Journal of Medicinal Chemistry. 2000;**43**:953

[11] Brockunier LL, Parmee ER, Ok HO, Candelore MR, Cascieri MA, Colwell LF, et al. Bioorganic & Medicinal Chemistry Letters. 2000;**10**:2111

[12] Chandrasekhar S, Prakash SJ, Jagadeshwar V, Narsihmulu CH. Tetrahedron Letters. 2000;**42**:5561

[13] Chandrasekhar S, Narsihmulu CH, Jagadeshwar V. Synlett. 2001;**5**:771

[14] Christensen C, Tornøe CW, Meldal M. The Journal of Organic Chemistry. 2002;**67**:3057

[15] Rostovtsev VV, Green LG, Fokin VV, Sharpless KB. Angewandte Chemie, International Edition. 2002;**41**:2596

[16] Tornøe CW, Christensen C, Meldal MJ. Organic Chemistry. 2002;**67**:3057

[17] Wang Q, Chan TR, Hilgraf R, Fokin VV, Sharpless KB, Finn MG. Journal of the American Chemical Society. 2003;**125**:3192

[18] Lober S, Rodriguez-Loaiza P, Gmeiner P. Organic Letters. 2003;**5**:17-73

[19] Scriven EFV, Turnbull K. Chemical Reviews. 1988;**88**:297

[20] Larhed M, Hallberg A. Drug Discovery Today. 2001;**6**:406

[21] Suzuki M, Kato N, Motomu Kanai M, Shibasaki M. Organic Letters. 2005;**7**:2527

[22] Sheng C, Zhang, Ji WH, Zhang M, Song Y, Xu H, Zhu J, et al. Journal of Medicinal Chemistry. 2006;**49**:2512

[23] Bioorganic & Medicinal Chemistry Letters. 2009;**19**(13):3611-3614

Chapter 5

Green Water Treatment for Pharmaceutical Pollution

Nilce Ortiz

Abstract

Environmental sustainability demands the advancement in water treatment and the use of lighting natural resources. Brazil has one of the most stable and intense solar irradiation in the world. It has to be used not only for energy generation purposes but also and mostly for water treatment, water quality polishment, and furthermore water disinfection. The chapter performs a comparison of different green technologies for water treatment as natural solar irradiation, simulated solar photolysis, solar photo-Fenton with and without hydrogen peroxide addition (solar/Fe), solar photo-Fenton with and without peroxide (solar/Fe/H_2O_2), titanium oxide-mediated photocatalysis (UV/TiO_2), photolysis under UV irradiation, and UV treatment with hydrogen peroxide (UV/H_2O_2). The chapter describes the solar photodecomposition calculations for pharmaceuticals and the emerging pollutants mostly found in polluted waters, including the decomposition route, kinetics, and process parameters. Many published works to point out the important properties to evaluate catalyzer and semiconductor materials after their use in photodecomposition processes. The essay includes the solar photodecomposition of dyes, carbamazepine, hormones, acetaminophen, antipyrine, bisphenol A, antibiotics, and the photodecomposition by-products. Finally, the chapter presents the synergistic effect between them with the probable mechanism and mineralization degree.

Keywords: pharmaceuticals, antibiotics, TiO_2, semiconductor, solar, photodecomposition

1. Introduction

The environmental performances are estimated using the life cycle assessment and the effective removal of 1 μg of 17α-ethynylestradiol (ES) for 1 liter of wastewater. The choice of ES was due to being a worldwide common micropollutant and endocrine-disrupting chemicals (EDCs) in a functional unit [1]. The natural solar photolysis exhibited an environmental footprint about 23 times greater than solar/Fe systems. The solar/Fe/H_2O_2 minimized the environmental footprint with the energy-intensive simulated solar irradiation that had a much higher, about five times, environmental footprint than the natural solar light. The UV photolysis also shows a low environmental impact. The TiO_2 addition to UV and H_2O_2 to UV reduced the environmental impact of about 97% and 88%, not considering the footprint of the TiO_2 production [2].

The total environmental footprint and the environmental sustainability of all the photodecomposition processes were directly proportional to water treatment efficiency and inversely proportional to treatment time (substantial energy input per time unit). The addition of TiO_2, iron, and H_2O_2 improved the process efficiency and environmental sustainability considering only the electricity consumption; the introduction of renewable energy resources could reduce the environmental footprint of the oxidation processes by up to 87.5%.

A solar spectrum contains approximately 46% of infrared light, 4% of UV light, and 50% of visible light. Such natural radiation composition highlight the importance of the visible-light-responsive TiO_2 development for natural and renewable solar photodecomposition applications. TiO_2 displays photocatalytic role when irradiated by ultraviolet (UV) due to its large bandgap (3.0 eV for rutile and 3.2 eV for anatase). The development of visible light-responsive TiO_2 includes chemical doping and photosensitization. The chemical doping is used to narrow the bandgap of TiO_2; the doped ions in TiO_2 act as recombination centers for photoexcited electrons and holes decreasing the photocatalytic activity.

The organic dye photosensitization on TiO_2 represents the major limitation for applications due to the poor dye stability, which can undergo desorption, photolysis and oxidative degradation, and fastback electron transfer, which results in low quantum yield for the photocatalytic reaction. An alternative for organic dye use in the metallic nanoparticles (NPs) is used successfully as photosensitizers for TiO_2 due to their stability and strong photoabsorption under visible light based on surface plasmon resonance [3]. It is a coherent oscillation of electrons on the metallic NP surfaces during the incident light irradiation.

The TiO_2 photosensitizers were first described in 2005 using gold (Au) as nanoparticles with TiO_2 (Au/TiO_2) the resultant film oxidized the ethanol and methanol at the expense of oxygen reduction under visible light radiation. The use of Au/TiO_2 for 2-propanol decomposition is induced by surface plasmon resonance photocatalytic process: (1) the Au NP adsorption of photons, (2) the Au NP electrons are injected into the conduction band of TiO_2, and (3) the resultant electron-deficient Au NPs can oxidize 2-propanol to recover to the fundamental metallic Au NP state. The pharmaceutical waste in the environment cause adverse health effects in the reproductive, neurological, and immune systems of wildlife and humans. Commonly they are resultant of agricultural lands' runoff, septic tanks' leakage, and the discharge of sewage treatment plants' effluents [4]. They can cause significant severe environmental damage even in trace concentrations. Worldwide EDCs are continuously released, and the efficiency of conventional water treatment technologies against such contaminants is minimal.

Advanced oxidation processes (AOPs) are a promising technology for the removal of the persistent organic pollutants (POPs), such as pharmaceuticals and endocrine disrupters (EDCs) from polluted waters. The photocatalytic methods are likely the most promising, especially those involving visible light-responsive materials, i.e., heterogeneous solar photocatalysis.

The photocatalytic hydroxide radicals are potent oxidants and react fast and unselectively with surrounding chemical species via radical addition, hydrogen abstraction, or e- transfer mechanisms. Many oxidant species can also degrade the pharmaceuticals and endocrine disrupter compounds in some cases the intermediate compounds present higher decomposition effectiveness and ends up in complete mineralization with the production of CO_2, H_2O, and inorganic salts.

The mass transfer is the controlled slowest step of the photodecomposition reaction of adsorption/desorption. The reaction requires the reactant adsorption/desorption on the catalyst surface and the presence of any barrier to reduce the reaction efficiency. Some experiments are performed keeping the reactor under dark conditions until the adsorption process reaches the equilibrium with the adsorbed mass at the semiconductor surface. After that, the reactor is irradiated and produces the radicals acting in the adsorbent surface and also in the solution. The radiation and the semiconductor generate e^- and h^+ pairs and radicals.

2. Photodecomposition: kinetics and process parameters

The heterogeneous photocatalysis shows a strong dependence on the operating temperature. The kinetics is usually dependent on the first step of the adsorption and the equilibrium modeled by Langmuir isotherms and Langmuir-Hinshelwood model. The first pseudo-order usually appears at the beginning of the reaction, just in the initial steps. As the reaction proceeds, the intermediate production could interfere with the radiation incidence. The observed decomposition rates as the pollutant start to decompose and begin the competition for the adsorption sites between the pollutants and the adsorbed species. The initial pollutant concentration starts to be the limiting reactant with mass transfer limitations with no kinetic control for lower concentrations.

The increase of the catalyst mass promotes the photodegradation rate and the active catalyst sites. If the system is a slurry, the reaction rate reaches a maximum or optimal value and after that declines. The higher suspended particle concentration enhances the light scattering, the particle agglomeration, with light opacity enhancement, and after a certain point, the photodecomposition efficiency decreases. There is an equilibrium between the available surface site and the suspended particles control.

The oxygen plays a vital role in the photodecomposition reactions; dissolved O_2 reacts with the photogenerated electrons leading to O_2^- radicals and preventing the recombination of the generated e^- and the h^+ pairs. The comparison between the kinetics of the air-saturated solutions and the pure oxygen solution often results in smaller rates for the first.

The pH values influence the catalyst aggregations and its surface charges, with the valence band (VB) and conduction band (CB) position. At low pH values, the CB holes are more effective in comparison with VB. The change in the pH values allows the surface charge modification mostly when the amphoteric groups are present [5].

The adsorption equilibrium after decomposition depends on the pollutant pH speciation and the reactive species. Considering the low pH values, the h^+ can be the more oxidizing species, the amount of $^{\cdot}OH$ increases under alkaline conditions when the OH^- ions are available to react with h^+, and the $^{\cdot}OH$ become the primary oxidant (Eqs. (1) and (2)). This effect increases the $^{\cdot}OH$ availability at higher pH values in spite of the negatively charged catalyst surface and also the pollutant repelling action at such pH values; the $^{\cdot}OH$ radicals' attack can explain the pH medium behavior.

$$(M - OH)\text{surface} + H^+ <=> (M - OH_2^+)\text{surface} \tag{1}$$

$$(M - OH)\text{surface} + OH^- <=> (M - O^-)\ \text{surface} + H_2O \tag{2}$$

The presence of some anions like chloride, nitrate, and sulfate reduces the photocatalytic performance as a result of the competition for the adsorption sites to scavenge the $^{\cdot}OH$ radicals. Natural organic matter and humic acids are scavengers of the reactive species and usually show negative photodegradation influence. Nevertheless, the presence of the carbonate and bicarbonate increases photodecomposition efficiency.

The temperature has a limited effect in the photodecomposition efficiency until 80 C°; after that there is the tendency to reduce the photocatalytic efficiency as a result of lower oxygen solubility in water. Different temperatures can also promote intermediaries and by-product formation.

3. Semiconductor materials

Semiconductor photocatalytic process has shown great potential as a low cost, environment-friendly treatment technology in degrading a wide range of pollutants. The photocatalysis has the dependency of the reactive oxygen species formation by the semiconductor particle with light energy greater than the bandgap energy [4]. The TiO_2 photocatalyst has an important drawback of photocatalysis and with gap energy which is the use of UV light, corresponding with 3–5% of natural solar light.

Some modification on TiO_2 surface is one promising route to enable TiO_2 sensitive to visible light for water purification. A variety of strategies improve the photocatalytic efficiency from dispersed solids to second-generation photocatalysts (chemically doped and physically modified by dispersed solids) achieving better spectral sensitivity and photoactivity.

Published results indicate better results with inert materials as zeolite (TiO_2-FeZ) or TiO_2 (SnS_2). Zeolite showed high surface area but lower bandgaps in comparison with TiO_2 powder and decreases the efficiency following the FeZ and SnS_2 content [5]. The H_2O_2 addition enhanced solar-driven degradation and solar/TiO_2-FeZ with higher decomposition rates followed by solar/TiO_2-SnS_2 with 51%, TiO_2 P25 with 41.3%, and finally 34.4% for TiO_2-SnS_2/H_2O_2. The pseudo-first-order kinetics was the driving for solar photodecomposition with the higher rates for solar/TiO_2-FeZ/H_2O_2 with $K_1 = 15.39\times10^{-3}$ min^{-1} with more than twice of the solar/TiO_2-SnS_2 rate and three times more than solar/TiO_2, solar/TiO_2/H_2O_2, and solar/TiO_2-SnS_2/H_2O_2.

A challenge to be overcome is the presence of suspended solid particles in the reaction environment; they reduce the solar irradiance and the photocatalysis efficiency. The solid deposition over an inert material with the high surface area can fix and stabilize the solid particles. The engineering materials with transitional metals in surface deposition are a solution, like carbon nanotubes, dye sensitizers, conductive polymers, graphene oxide, and other semiconducting materials. The semiconductor supporting material has to avoid the agglomeration formation.

The commercial TiO_2 (P25) is the most common catalyzer in spite of lower azo decomposition; Ag_2O is another active photocatalyst with promising results for azo photodecomposition. Nevertheless, the use of Ag_2O in azo mixture showed faster degradation with better decomposition results due to the synergetic oxidation effect.

The scavengers' presence reduces the photodecomposition effect in water suspension. The ions HCO_3^-/CO_3^{2-}, SO_4^{2-}, Cl^-, and NO_3^- showed inhibitory effects toward the hydroxyl radicals generated by AOPs.

The natural organic matter (NOM) presence showed a synergistic effect increasing the E2 degradation, such degradation produces other radicals. Published studies relate to the degradation of the NOM species of humic acid and fulvic acid applied in solar/photodegradation resulting in all organic compounds that were mineralized after 150 min of treatment. The formed E2 intermediaries during the treatment do not possess any estrogen effect.

The addition of noble metals as Au and Ag increases the visible light ability with the manipulation of the optical properties and microstructure combined with inorganic and biotemplates as nanostructures of micelles, as spent tea leaf, trimethylammonium bromide, and metal nanoparticles improve the photodegradation efficiency. The use of Co_3O_4 spinel nanoparticles, NiO nano-sticks, the binary metal oxide nanocomposites of CeO_2/Y_2O_3 and NiO/MnO is effective in dye degradation of RhB, MB, MO, and rose bengal dye.

BiOCl has superior efficiency as photocatalysts due to the interlayer-specific structure of $[Bi_2O_2]_2^+$ with double Cl^- ions where the photogenerated e^- and h^+ pairs are separated. The BiOCl microspheres synthesized via ethylene glycol are mediated by a solvothermal method with the visible light drive. The crystallinity, surface area, and optical and electronic properties of BiOCl samples depend on the reactant concentration with the benefit from the exposed (110) face and oxygen vacancy; BiOCl allows a maximum CBZ degradation efficiency of 70% after 180 min under visible light illumination. The kinetic rate constant (k) of CBZ degradation in synthetic BiOCl (0.0935 min^{-1}) was 52 times higher than the ordinary BiOCl (0.0018 min^{-1}). The improved photocatalytic activities for BiOCl were attributed to the combination of enhanced carbamazepine adsorption, increased with visible light drive and efficient separation of photogenerated e^- and h^+ pairs. The trapping experiments of radicals and holes showed the $^{\cdot}O_2^-$ and h^+ as dominant active species in the process and the most important; the BiOCl performance was also efficient in natural water without any additive. The experimental findings indicate the BiOCl photocatalysis is an efficient and cost-effective technology for recalcitrant pharmaceutical contaminant removal.

4. Dye photodecomposition

The higher development of the textile industry caused the emission of large quantities of dye wastewater with high chemical stability in surface water resources all over the world; the effect is the severe environmental damage and problems worldwide.

The advanced oxidation process (AOP) is in situ treatment technology and is widely applied on persistent, toxic, and poorly biodegradable organic pollutants.

The improvement of photodecomposition process reduces the by-products' and final products' toxicity. The biological methods are insufficient to decompose such stable organic compounds and chemical molecules. Industrial wastewater is a mixture of various components with high complexity and diversity. The interactions among the different components can occur, weakening and even blocking the photodecomposition effect. The heterogeneous photocatalysis is taking considerable attention to the textile wastewater treatment due to its low cost and low secondary by-product pollution. The disadvantages are the low quantum efficiency and slow reactant rate using the most common semiconductor, the TiO_2. The use of Ag_2O with a very narrow bandgap of 1.3 eV allows applying a wide range of the solar spectrum with an increase in the photodecomposition rate. The literature describes a photodecomposition process with last about 120 s to degrade the Methyl orange under UV and Visible light and 40 min with only Infrared light.

The application of the visible light photodecomposition in a dye mixture of methylene blue (MB), methyl orange (MO), and rhodamine (RH) indicates the MO as the more stable azo compound than the other organic pollutants due to the aromatic groups attached at the end of the azo bond. Despite this fact when the light-driven photodecomposition uses Ag_2O as a catalyzer, it was the fastest and easiest decomposed compound. Published results indicate the visible light photodecomposition with Ag_2O with the elimination of 90.2% of MO, 96.5% of RH, and 99.5% of MB using 4, 50, and 20 min, respectively.

In dye photodecomposition in acidic conditions, some peaks with higher absorbances change some wavelength numbers indicating the chemical structure transformation from hydrazone to azo form. Despite such change, the concentration still reduces with time, and the complete degradation of the dye mixture finished in 18 min at pH 3 and 15 min at pH 5. The observation of 90% of the total mineralization was after 50 min under acidic conditions and 40 min under neutral and alkaline conditions.

The dye mixture showed better decomposition results than only a single one; a synergistic oxidation phenomenon occurring in the photodegradation of the dye mixtures with Ag_2O indicates no apparent photoreduction of Ag_2O, and the solid material still consisted of pure Ag_2O and can be used consequently as a high-performance catalyst for dye wastewater treatment.

5. Solar treatment for carbamazepine

The carbamazepine (CBZ) is widely used as antiepileptic and mood stabilizer, worldwide, and the consumption is about 1014 ton $year^{-1}$. The CBZ shows high stability and low biodegradability, and the removal percentage for conventional water treatment process is less than 10%. Almost all consumed CBZ is discharged as sewage in the water environment causing adverse effects on the surface water quality, ecosystem, and human health [6].

The CBZ photodegradation process catalyzed by TiO_2 and ZnO nanoparticles generates three derivatives: carbamazepine epoxide, acridine, and acridone. TiO2 is effective in degrading CBZ and carbamazepine epoxide. Considering the acridine and acridone, no significant differences were found between those two catalysts. The CBZ and carbamazepine epoxide photodegradation was affected by pH (especially in the presence of TiO_2 as NPs) and natural organic matter [7]. In contrast, the acridine and acridone photodegradation was not affected by pH and organic matter. The TiO_2 and ZnO catalysts present contrasted efficiency on CBZ decomposition when compared with its derivatives and the effect of environmental parameters on the CBZ as photodegradation efficiency of the derivatives' presence cannot be predicted based only on the parent molecule (CBZ) behavior. The indication of higher degradation efficiency was for higher initial concentrations with a degradation rate of 52 times greater than for lower initial concentrations, and the kinetics corresponds to pseudo-second-order.

The negatively charged CBZ is due to the presence of the amide bond, and the exposed surface of the semiconductor positively charged enhances the surface adsorption processes. The proton adsorption also improves the effectiveness of the catalyzed reaction under visible light radiation, promoting long-term stability and the catalyst reusability. The visible-light-driven photocatalytic activity enhancements are the synergetic effects, including a large absorption capacity, the promotion of the light harvestability, and high separation efficiency of photogenerated e^- and h^+ pairs.

The use of some scavengers as $AgNO_3^-$ for e^-, HCOONa for h^+, and butanol for $^{\cdot}OH$ (radical) and the N_2 purging to detect the function of O2 elucidate the main active species involved in the photodegradation process. The addition of $NaHCO_2$ reduces the decomposition in 18% as an indication of the importance of the h^+ radical in the reaction. The addition of n-butanol or N_2 purging showed no different reaction efficiency; this effect implied the addition of O_2 produces the $^{\cdot}O_2$ radicals and increases the CBZ degradation confirming the domination of the degradation process by oxidation step with the $^{\cdot}O_2^-$ radical generation and partly by the direct h^+ oxidation process.

The CBZ (A) oxidation reaction results in four intermediaries B, C, and D. The compounds B, C, and D are intermediaries with high decomposition rate (**Figure 1**). The attack of the CBZ oleofinic bond in a central heterocyclic ring forms compound B. The generation of radicals and the oleofinic bond of CBZ decomposed in C and D. The compound C uses a ring contraction reaction followed by an intramolecular cyclization mechanism resulting in intermediate compound F. Finally the formation of compound G was by intramolecular cyclization mechanism of compound D.

The primary concern about the pharmaceutical photodecomposition is the possible formation of the toxic intermediate and by-products. The literature cited as possible toxic intermediates for CBZ photodecomposition, the CBZ-10, 11-epoxide,

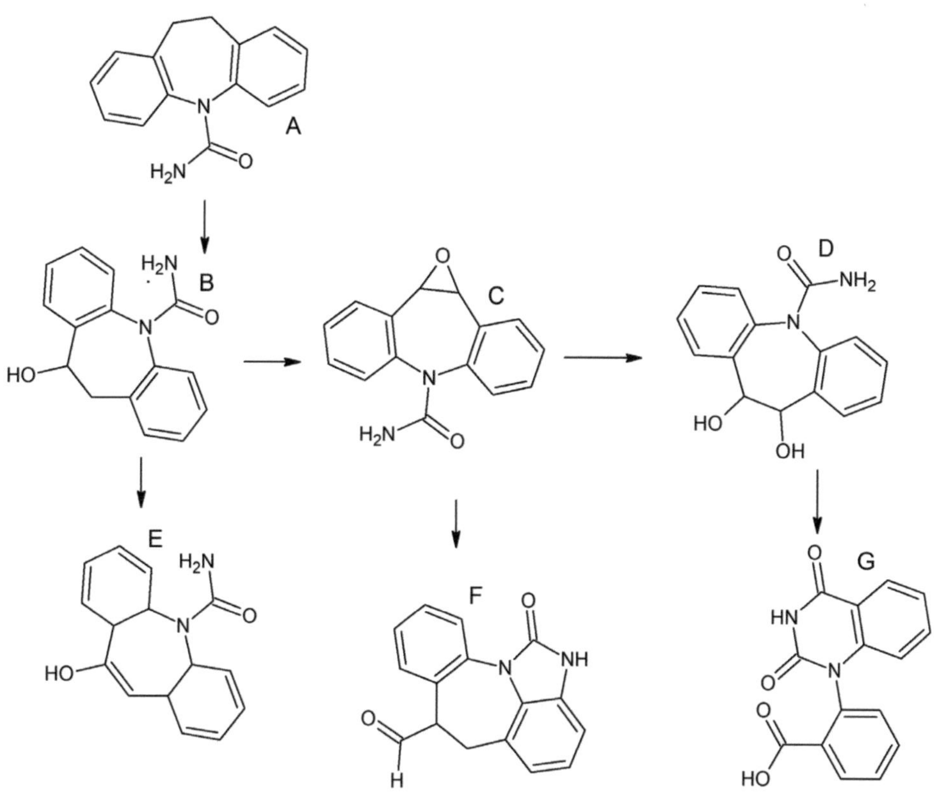

Figure 1.
The Carbamazepine degradation by-products.

and acridine. The photodecomposition optimization indicates BiOCl removed about 2–3 mg L^{-1} of CBZ after 150 min of solar irradiation, more than 78% of no toxic intermediaries. Always there will be the possibility to keep the photodecomposition reaction for a longer time of solar radiation. The recommendation is the use of adsorbent material, and after the absorption, removing the pharmaceuticals from the reaction media, and the photodecomposition reaction can keep continuing until mineralization and proper discharge.

6. Solar photodecomposition of hormones

The 17β-estradiol (E2) is the most natural estrogenic hormone occurring in sewage-polluted waters and also an intermediate key in the industrial synthesis of other estrogens. It is frequent in natural water environment with the high potential to hormonal disruption pathways in wildlife even in low nanogram concentrations. Recently, it was added to the watch list of priority substances in the EU Water Framework Directive. Many research projects use E2 as representative of emerging pollutant (EP) for water tertiary treatment study and photodecomposition improvement [8].

There are many studies of immobilized TiO_2-based composites, TiO_2- and iron-exchanged zeolite of ZSM5 type (TiO_2-FeZ), or another semiconducting material (TiO_2-SnS_2) and active solar photocatalysts. The solar-driven photocatalytic parameters as pH values, H_2O_2 concentration, and composite formulation, on the effectiveness of E2 degradation, allow the calculation of the surface modeling. The solar/TiO_2-FeZ/H_2O_2 process achieved E2 degradation by 78.1%; it was higher in comparison with the reference process of TiO_2 P25 with 41.3% of remotions and the solar/TiO_2-SnS_2 and solar/TiO_2-SnS_2/H_2O_2 processes with 51.0 and 34.4%, respectively. The E2 degradation by solar/TiO^{2-}FeZ/H_2O_2 enhances in the presence of NOM, as real water constituents. On the other hand, the nitrates and carbonates presence show an inhibitory effect.

There are many studies using visible photodecomposition of the hormones such as estrone (E1), 17β-estradiol (E2), estriol (E3), and 17β-ethinylestradiol (EE2) with the concentration in the interval of 0.004–5.00 mg L^{-1} using TiO_2 in porous sheets, microcrystalline glass plates, P25 suspension, UVC/H_2O_2, solar Fe II, and ozone 30 mg L^{-1}. The process usually includes UV and visible light from 280 to 400 nm and full spectrum from 200 nm to 30 μm, and LED lamp with the lines at 382 nm, 254 nm, and 254 high intensity. The kinetics of such process was from 550×10^{-3} min^{-1} (pH 5) to 3.4 min^{-1} (20°C) with and without H_2O_2 with solar light 86% in 60 min and 63.9% in 12 min. The hormones enter in the water environment mainly from the sewage discharge and the effluent of sewage treatment plants. The EE2 is the main composition of the oral contraceptives, and E1, E2, and E3 occur naturally. The estrogenicity order is EE2 > E2 > > E1>>>> > E3. Most of the hormones show photodegradation over immobilized TiO_2 sheets under UV-LED irradiation or solar radiation following the first-order kinetics, faster at pH 4.

The most efficient hormone decomposition is the combination of the photodecomposition and ozonation using TiO_2-coated glass with LED irradiation on $\lambda = 382$ nm. The application of such a dynamic process removes about 22 chemical priority substances and contaminants of emerging concern, including the resistant bacteria and genes after discharge in surface water resources. There was no compound with estrogen effect formed after the reaction; the process improves the removal efficiency of microbial loads.

The improvement of the EE2 degradation, clofibric acid, nonylphenol, and carbamazepine was slightly by the use of ultrasound combined with ozonation and photodecomposition. The removal percentage increases with pH, but at higher pH also showed an adverse decomposition effect.

The test in vitro can detect estrogen receptor agonists and antagonists. However, the estrogen-disrupting compounds do not only act on the estrogen receptor but also inhibit enzymatic catalysis reactions and the transport of hormones in the blood or the hormone production. Only in vivo analysis with a full spectrum of possible mechanisms can be identified in a whole organism.

The photocatalytic degradation of 17α-ethynylestradiol (EE2) allows the identification of 12 intermediates. The decomposition efficiency of EE2 decreased at pH 3 and in the presence of methanol at pH 7. The study proposes three degradation pathways: (1) the transformation of the phenolic ring, (2) the photocatalytic degradation of the aliphatic carbon linked to the aromatic ring at pH 7, and (3) the isomerization of EE2 in the presence of methanol at pH 7. The EE2 photocatalytic degradation is pH-dependent and at pH of 3, 7, and 10 without methanol addition were 63, 72, and 99%, respectively. The pH increase facilitated the formation of hydroxyl or hydroperoxyl disubstituted intermediates. In aqueous solutions, aliphatic carboxylic acid decarboxylation is preferable to the corresponding reduced hydrocarbons; some published results indicate the favored pathway for dicarboxylic acid mineralization is the decarboxylation resulting from the photo-Kolbe reaction [Eq. (3)]. The responsible for the formation of intermediary compounds is the attack of the $^{\cdot}OH$ and $^{\cdot}OOH$ radicals:

$$RCOO^- + h^+ \rightarrow R^{\cdot} + CO_2 \text{ (photo - Kolbe)} \quad (3)$$

First was the oxidation of the EE2 and then the reduction to EEO in the presence of h^+ and electron, formed by the methanol radical as described in [Eqs. (4)

and (5)]. Under acidic conditions, there is the inhibition of the radical methanol production. Alternatively, under alkaline condition, very little free h^+ existed in the solution, and the formation of EEO only occurred under neutral pH conditions:

$$h^+ + CH_3OH \rightarrow CH_3O^{\cdot} + H^+ \quad (4)$$

$$CH_3O^{\cdot} \rightarrow H_2CO + H^+ + e^- \quad (5)$$

The photodecomposition of methanol acts as an $^{\cdot}OH$ scavenger and retards the photocatalytic degradation of EE2 reducing for only 8, 11, and 15% at pH of 3, 7, and 10, respectively, with methanol presence. Published researches indicate just the addition of a small amount of methanol or toluene inhibited the photocatalytic oxidation.

The formation of the intermediate products was in low mineralization rate and low removal efficiencies, resulting in total organic carbon less than 20% with the occurrence only of the phenolic ring transformation during the reaction.

7. Solar treatment for acetaminophen and antipyrine contamination

The acetaminophen (ACE) is one of the most widely used analgesics and antipyretic drugs and is one of the top pharmaceuticals prescribed in the USA or England, being China the second ACE manufacturer. It is present in surface water bodies as a result of 60–70% of human excretion via urine after medicine consumption [9]. The ACE water detection in the USA and Europe on sewage treatment plant and their effluents was in concentrations of 10 to 65 μg L^{-1}. In population, such water pollution may lead to hepatic necrosis caused by its transformation to N-acetyl-benzoquinone imine upon oxidation, which can hydrolyze to 1,4-benzoquinone; both by-products are toxic of significant concern.

The antipyrine (ANT) detection is common in sewage and polluted surface water. Such anti-inflammatory compound is a nonsteroidal and antipyretic drug which enters in the aquatic environment after use. Environmental accumulation causes adverse human health effects and affects aquatic life. The concentrations of emerging contaminants in the influent and effluent from wastewater treatment plants showed the concentration of ANT was relatively low (about 0.04 mg L^{-1}), and about 68.5% escaped from conventional activated sludge wastewater treatments, allowing to reach surface water resources.

TiO_2 is still the widest semiconductor used for ACE and ANT photodegradation due to its low cost, nontoxicity, and chemical stability. Nevertheless, the difficulty of TiO_2 recovery after the reaction and the relatively limited adsorption capacity with low surface area and porosity are some of the technological disadvantages.

There is no acetaminophen degradation versus time under solar irradiation in the absence of photocatalyst after 6 h; the removal using different photocatalysts without light irradiation also can be considered negligible (lower than 6%). The sensitization of TiO_2 by the carbon material titanium nanotubes and C-Ti showed a significantly higher activity than the non-modified Ti. However, the acetaminophen removal remained below 70% after 4 h of illumination. The lack of anatase crystal structure is responsible for the small photoactivity; the amorphous titanium is not active. The air calcination at above 300°C had a beneficial effect on C-Ti catalyst; the calcined samples at 400 and 500°C allowed the total acetaminophen conversion in only 1 h. The crystallization of TiO2 explains the effect into anatase (with a band-gap energy of 3.12 eV), which is the most active titanium phase for photocatalytic applications; at the lowest calcination temperature tested (300°C), there is no significant crystallization of anatase.

Published results indicated the pseudo-first-order rates were 0.13, 0.19, and 0.38 h^{-1} for complete photodecomposition of antipyrine, acetaminophen, and ibuprofen, respectively. Regarding with the properties of the C-TiO_2 semiconductor materials, the structured defects caused by the C incorporation (as substitutional anion or interstitial cation) are the responsible for the photocatalytic activity of these materials, acting as trap centers or the photogenerated charges.

The investigation of the role of the reactive oxygen species used selected scavengers as isopropanol, the OH radical scavenger; the addition reduced the degradation rate. The OH radicals are very reactive, and the reduction by the scavenger inhibited the degradation rate, an indication of the involvement of the OH radical production in acetaminophen photodegradation. Some references mentioned the O_2^- radicals attack preferentially organic compounds with aromatic rings (as ACE aromatic ring).

8. Photodecomposition of bisphenol a (BPA)

BPA is present in polycarbonate plastics and epoxy resins such as plastic water bottles, many food containers, water pipes, medical equipment, dental sealants, thermal receipts, electronics, and toys [10, 11]. The compound is toxic for the reproductive system because it mimics the human hormone estrogen. BPA production in the world exceeds 3 million tons per year. The BPA presence in the environment results in adverse effects on organic metabolism such as reproduction, metabolic systems, organism development, neural networks, and cardiovascular irrigation. The use of BPA is mostly as a plastic monomer, the monomer for epoxy, polycarbonate plastics, and epoxy resins. About all studied environment compartments have in some degree any content of BPA including air, water, and soil [9]. Published works indicate a connection between the BPA exposition and high levels of anxiety, depression, hyperactivity, and inattention. The BPA detection in organic body demonstrated its presence in blood, urine, cardiovascular diseases, diabetes, and obesity, posing a risk for fetal development and reducing the basal testosterone secretion. There is also a combination of BPA presence and other similar compounds in environmental compartments, food and food containers, and also in humans' milk, urine, and placental tissue; this is evidence of the possible global exposition.

The use of spiked sodium hypochlorite removes BPA from real water samples at 50 mg L^{-1} for 10 min with a removal percentage of 99%. In spite of the formation of chlorinated by-products during the process with some toxic side effects. The advanced UV/H_2O_2 was able to remove 85% of the initial BPA at 240 min. However, a high level of the H_2O_2 is essential to execute such BPA removal process. The presence of carbonates and bicarbonates reduces the UV/H_2O_2 efficiency due to scavenging radical's formation. The ozonation is an excellent option but is extremely costly and is suspected to form intermediates which carcinogenic nature [12].

The concentration = 20 mg L^{-1}, TiO_2 dosage = 0.5 g L^{-1}, initial pH = 7.0, and temperature = 25°C followed the first-order model. The possible mechanisms for BPA photodegradation are in the following sequence:

1. Initial photooxidation, proceeded by electrophilic hydroxyl radicals ($^{\cdot}OH$), produced the photocleavage of electron-rich carbons in the phenyl groups of BPA or the excited BPA molecules attacked by hydroxyl radicals ($^{\cdot}OH$) forming phenol radicals ($^{\cdot}C_6H_4OH$) and isopropylphenol radicals ($^{\cdot}C(CH_3)_2C_6H_4OH$) [Eq. (6)].

$$HOC_6H_4C(CH_3)_2C_6H_4OH + (^{\cdot}OH) \rightarrow {}^{\cdot}C_6H_4OH + {}^{\cdot}C(CH_3)_2C_6H_4OH \quad (6)$$

2. The hydroxyl radical ($^{\cdot}OH$) converted to p-hydroquinone (HOC_6H_4OH) and isopropylphenol radical allow the formation of the 4-hydroxyphenyl intermediates such as p-hydroxybenzaldehyde, p-hydroxyacetophenone, and 4- hydroxyphenyl-2-propanol [Eq. (7)].

$$\begin{aligned}&{}^{\cdot}C_6H_4OH + {}^{\cdot}OH \rightarrow HOC_6H_4OH + {}^{\cdot}C(CH_3)_2C_6H_4OH + {}^{\cdot}OH \rightarrow \\ &HOCC_6H_4OH \text{ or } (HOC(CH_3)C_6H_4OH \text{ or } HOC(CH_3)_2C_6H_4OH) \quad (7)\end{aligned}$$

3. The oxidation reaction of the single-aromatic intermediates through ring-opening reactions results in aliphatic acids.

The pH values decrease in aqueous media gradually, and the intermediates were entirely mineralized of forming carbon dioxide (CO_2); the oxidization of the aromatic intermediates occurs subsequently through ring-opening reactions into aliphatic acids [Eq. (8)].

$$\begin{aligned}&\text{Single - aromatic intermediates} + {}^{\cdot}OH \rightarrow \text{aromatic ring opened products} \\ &+ {}^{\cdot}OH \rightarrow CH_3COOH + {}^{\cdot}OH \rightarrow CO_2 + H_2O \quad (8)\end{aligned}$$

The UV-A radiation (λ = 365 nm) with TiO_2 P25 result in complete removal after 180 min in first-order kinetics with Kap = 20.3 10^{-3} min^{-1}, with 5 mgL^{-1} as initial concentration and 200 mg L^{-1} as the addition of the TiO2 P25 [3]. The complete mineralization of BPA was at pH 3 after 120 min; intermediates formed at higher pH values are most stable and therefore difficult to be decomposed and mineralized [13, 14]. The TiO_2 sources as anatase, rutile, brookite, and their mixtures indicate the lower uptakes with less than 6% were over the raw TiO_2. The anatase and rutile TiO_2 mixtures obtain 94 and 80% of removal percentage, respectively, higher than the obtained for the anatase, rutile, and brookite single composition. The mixture anatase/TiO_2 brookite reached the complete mineralization, and the mixture of anatase and rutile was five-fold slower than the commercial TiO_2 P25 with 3 min of complete removal percentage. All products showed less toxicity and estrogenic activity than the initial BPA [15, 16].

The nano-TiO_2 facilitates the degradation under sunlight radiation with O_2^{-2} as dominant oxidizing species. The better degradation efficiency was at pH 2.6, and correspondent with the pseudo-first-order without nanoparticles was one or two orders small with λ = 365 nm of radiation using pristine nanotubes [10].

The higher results were with anatase particles enhanced by the presence of rutile and preferential oxidation of reaction intermediates on brookite. The toxicity removal was for TiO_2 supported on a glass fiber with UV light radiation λ = 365 nm in batch and stirrer tank.

The use of a wide range of metals as the lanthanum-doped TiO_2 was able to degrade BPA completely under acidic conditions within 2 h; the result is far better than undoped TiO_2.

The oxidants' addition enhances efficiency, as H_2O_2 and FeII. The H_2O_2 interacted with Fe-2þ ions to produce hydroxyl radicals. The Fe doped into the TiO_2 decreased the bandgap, which also enhanced the BPA photodegradation. The addition of 5 mol% of Fe in TiO_2 successfully removed 10 ppm of BPA in 2 h. The experiment with nitrogen doping on TiO_2 indicated the N-doped TiO_2 enhanced the photodegradation of BPA compared to conventional TiO_2. Likewise, the iodine-doped TiO_2, upon exposure to UV and visible irradiation, showed increased degradation efficiencies for BPA up to 93 and 100%, respectively.

The 4-chlorophenol, phenol, methylene blue, rhodamine B, and acid orange presence reduces the surface area for the volume and enhances the oxygen vacancies in TiO_2 surface matrix by N-doping and F-doping, and surface acidity is also improved by F-doping, and visible light adsorption by N-doping of nitrogen-fluorine-codoped TiO_2 photocatalyst. The use of simulated sunlight lamps promotes the generation of the active species for BPA decomposition as O_2^{-2}.

The production of TiO_2 PEG started with the mixture of titanium ethoxide and ethanol solution followed by PEG addition; the suspension aged for 24 h and calcined for 2 h at 400°C. The optimization of the TiO_2 production by sol–gel polyethylene glycol (PEG) includes the variation of the PEG molecular weight, the mass percentage, the pH, and the TiO_2 dose. The visible light BPA degradation rates for TiO_2 with PEG200 (10%), PEG600 (5%), and PEG3500 (0.5%) at pH 4 were 2.07, 3.01, and 2.90 h^{-1}, respectively. After 12 h of reaction, the total organic carbon measurements indicated a small BPA degradation with the reduction of TiO_2, with PEG200 (10%), PEG600 (5%), and PEG3500 (0.5%) of 38%, 56%, 65%, and 64%, respectively. The content of hydroxyl radicals in TiO_2, with PEG200 (10%), PEG600 (5%), and PEG3500 (0.5%), was 50.1, 88.6, 78.8, and 75.1 μM, respectively. Allowing the conclusion about the PEG addition on the TiO_2 preparation increases the photoactivity, and the optimal PEG addition percentage varied with PEG molecular weight and content.

9. Antibiotic photodecomposition

The antibiotic removal and other anthropogenic compounds by adsorption are the major chemical process of deactivation, and it is important to reduce the toxic properties and to restrict their transport into water systems. The adsorbent material in combination with titanium dioxide or titania (Ti) showed better results using adsorption combined with photocatalytic activity with low cost, nontoxicity, and high stability in aqueous solution. Nevertheless, the disadvantages of TiO_2 powders are the low surface area (Degussa P25 = 35–45 $m^2\,g^{-1}$, anatase < $m^2\,g^{-1}$); the anatase bandgap of 3.20 eV uses only a small UV fraction of solar light, about 2–3%, with the high cost of the TiO_2 powder separation and recovery from treated wastewater [17, 18].

The removal of tetracycline (TC) by TiO_2 and the mesoporous binary system TiO_2-SiO_2 was tested, and it is strongly dependent on pH, with increasing pH it decreases. The electrostatic forces and H-bond formations mainly between amide, carboxylic, and phenolic groups of the antibiotic and the functional groups of TiO_2 are also important. The adsorption capacity increases in the following order TiO_2 < TiO_2-SiO_2 (high surface area). The photodegradation rate is affected by pH 7 or lower; the related mechanism is to $OH^{\cdot}$ radicals—the composed titania-silica act as an adsorbent and alternative photocatalyst for pollution control.

All processes result in high degradation efficiency of the β-lactam antibiotic (oxacillin). The TiO2 photocatalysis, the sonochemistry, the photo-Fenton process, and electrochemistry (with a Ti/IrO2 anode in sodium chloride solution). The processes are successful ,but three of them involve the hydroxyl radical generation and the degradation pathways, by-products' generation, and the mineralization degree. The electrochemical process performed the decomposition by chlorine production and its attack when the sonochemical and photo-Fenton systems have the production of the hydroxyl radical.

The high oxidant species with low selectivities, such as hydroxyl radicals (E = 2.8 V), are formed in advanced oxidation processes (AOPs) and the active chlorine electrogenerated through dimensionally stable anodes (DSA). The irradiation of an aqueous suspension of TiO_2-semiconductor with UV light produces hydroxyl radical.

The TiO_2 photocatalysis combines the holes' generation with the attack of the hydroxyl radicals. The by-product analysis indicated the four oxidation processes exhibited the oxidation of the thioether radical followed by the amide breakdown and finally the β-lactam opening ring. However, the antibiotic decarboxylation was only a result of the TiO_2 photocatalysis, explained by the holes' production with direct oxacillin oxidation [Eqs. (9)–(15)]. The electrochemical process promotes the oxacillin isomerization pathway, while the photo-Fenton and TiO_2 photocatalysis treatments showed hydroxylation at the aromatic ring. The different degradation routes generated different mineralization extent and efficiency [19].

$$P + {}^{\cdot}OH \rightarrow Pox \tag{9}$$

$$HClO, CLO^{-}, Cl_2 + P \rightarrow Pox \tag{10}$$

$$TiO_2 + h\nu\ (\lambda < 387\ nm) \rightarrow TiO_2\left(e^{-} + h^{+}\right) \tag{11}$$

$$h^{+} + H_2O \rightarrow H^{+} + {}^{\cdot}OH \tag{12}$$

$$h^{+} + OH^{-} \rightarrow {}^{\cdot}OH \tag{13}$$

$$h^{+} + P \rightarrow P_{ox} \tag{14}$$

$$e^{-} + O_2 \rightarrow {}^{\cdot}O^{-}{}_2 \rightarrow \rightarrow {}^{\cdot}OH, {}^{\cdot}OOH, H_2O_2 \tag{15}$$

The total organic carbon measurements in TiO_2 photocatalysis and the photo-Fenton system were 90 and 35%, respectively, and with just the sonochemical and electrochemical treatments, the pollutant was not mineralized.

The presence of the ultrasonic waves in aqueous solutions is another way to form hydroxyl radicals [Eqs. (16)–(19)]. Singular conditions of temperature (5000 K) and pressure (1000 atm) induce the formation of ultrasonic microbubbles which violently collapse in water, and the dissolved oxygen is dissociated to produce hydroxyl radicals.

$$H_2O +))) \rightarrow {}^{\cdot}H + {}^{\cdot}OH \tag{16}$$

$$O_2 +))) \rightarrow 2{}^{\cdot}O \tag{17}$$

$$H_2O + {}^{\cdot}O \rightarrow 2{}^{\cdot}OH \tag{18}$$

$$O_2 + {}^{\cdot}H \rightarrow {}^{\cdot}O + {}^{\cdot}OH \tag{19}$$

The reaction of Fe (II) with hydrogen peroxide produces radicals. The reduction of Fe (III) in aqueous media results in Fe (II) by the action of UV–Vis light and extra hydroxyl radicals in a photo-Fenton process.

The electrochemical oxidation which soluble chloride results in chloride anions on the Ti/IrO_2 anode [Eq. (20)] which the generation of hypochlorous and hydrochloric acids [Eq. (21)]. The dissociation of the hypochlorous acid forms hypochlorite [Eq. (22)]. Chlorine, hypochlorous acid, and hypochlorite are active species, and they are very dependent on the pH values. The predominant species at pH lower than 3 is Cl_2 (E = 1.36 V), in the range of pH 3 to 8 is $HClO^-$ (E = 1.49 V), and at pH higher than 8 is OCl^- (E = 0.89 V).

$$2Cl^- \rightarrow Cl_2 + 2e^- \tag{20}$$

$$Cl_2 + H_2O \rightarrow HClO + HCl \tag{21}$$

$$HClO^- + H_2O \rightarrow ClO^- + H_3O^+ \tag{22}$$

The knowledge of the oxidation routes in water treatment can optimize the process and establish a pollutant degradation mechanism and pathways: the experimental parameters and the matrix influence on oxacillin (OXA) on electrochemical oxidation and TiO_2 photocatalysis. Here is no report about photo-Fenton and sonochemical processes' removal of oxacillin from polluted water.

The sonochemical process degraded the antibiotic and generates solutions without OXA entirely; the antimicrobial activity showed an excellent performance and adjustment to exponential kinetic-type decay, and the degradation rates were 1.4 μM min^{-1} for OXA, 1.3 μM min^{-1} for OXA with mannitol, and 1.4 μM min^{-1} for OXA with calcium carbonate. The possible OXA sonic degradation mechanism was proposed based on the evolution of the by-products and their chemical structure [Eqs. (16) and (17)] [20].

The ultrasound application over 120 min removed OXA compounds and eliminated its antimicrobial activity. However, the mineralization was not reached even after (360 min). The mineralization of the oxacillin under previous water sonication reduce the microbial activity even with non-adapted microorganisms from a municipal wastewater treatment plant. The results showed the sonochemical transformation of the initial pollutant into biotreatable substances even using the typical aerobic biological system.

The iron ions present in the matrix affect the antibiotic (OXA) decomposition, with improvement in degradation, and the inhibition was by the addition of pharmaceutical excipients of a commercial formulation or by inorganic ions of natural mineral water. The best performances were achieved at natural pH 6.0 using 2.0 g L^{-1} of TiO_2 with 150 W of light intensity. The OXA photodegradation process showed a Langmuir-Hinshelwood kinetic model. The achievement of the total antibiotic removal was after 120 min, with 100% of mineralization. Finally, the identification of five by-products elucidates the degradation routes with a proposition of an antibiotic degradation (**Figure 2**).

The addition of 2-propanol as a scavenger, 25 times higher than the antibiotic, produces a slight reduction (about 3%) in the antibiotic removal rate, and the concentration of 645 times higher than the OXA causes 30% of inhibition. The result indicates the hydroxyl radicals present at the solution may contribute to the degradation of the antibiotic molecules. The essays in the presence of KI concentration 25 times higher than the OXA concentration showed 75% of inhibition. The use of equimolar KI concentration resulted in a 13% reduction. The indication of the degradation rate is in association with adsorption reduction of the catalyst surface. Consequently, the degradation of OXA by heterogeneous photocatalysis seems to occur mostly at the catalyst surface and via two routes: by the radical attack and photo-Kolbe mechanism.

The UV irradiation of antibiotic molecules generates excited states and the detection of such reactive species by an indication of their ability to oxidize luminal reagent. Such compound uses the electronically excited aminophthalate, which

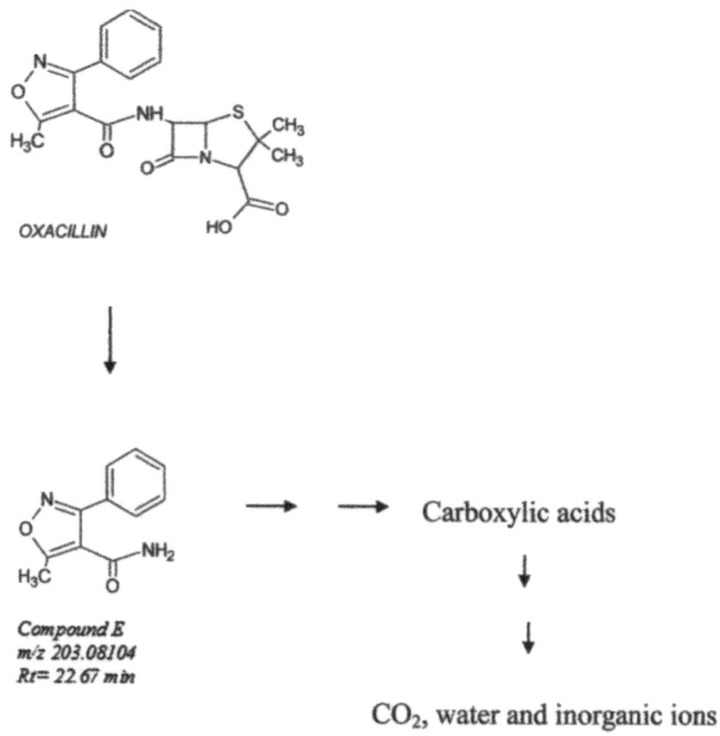

Figure 2.
Oxacillin photodegradation pathway.

decays to the ground state releasing electromagnetic radiation in the visible zone of the spectrum—the application of the method to penicillin G, nafcillin, azlocillin, and neomycin dissolved in water. The intensity of the luminal chemiluminescence emission (CL) was proportional to the radical concentration and dependent on the molecular structure of the drugs. Under the optimized conditions, the penicillin and azlocillin were the most susceptible to photodegradation, while neomycin sulfate was less affected by the UV light. The addition of a hydroalcoholic extract of rose petals to antibiotic solutions reduced the CL intensity, indicating the alcohol act as a scavenger of free radicals of the irradiated drugs.

10. Synergistic effect

In the application of the solar photodecomposition in the dye mixture of RH and MB, the result is similar with a single dye, and the adsorption balance remains unchanged with no interaction between RH and MB and their by-products. Nevertheless, the addition of MO in the mixture accelerated the photodecomposition significantly. The decomposition of RH/MO and the MB/MO reduced the decomposition time in 13 and 10 min, respectively [21]. Such an effect is positively dependent on the MO concentration; the application of Eq. (23) to the Langmuir-Hinshelwood model

$$-\ln(C/C_0) = k' Kt = kt \tag{23}$$

where C_0 and C are the initial and t measured concentrations, t is the reaction time, and k are the first-order-kinetic reaction constant, calculated using $\log(C/C_0)$ vs. t.

The kinetics k rate indicates higher values for binary systems with MO component. When the MO concentration reaches a constant value, the reaction depends on the photocatalyst mass. The preparation of ternary mixtures with RH or MB and different azo species as orange G (OG), methyl red (MR), and Eriochrome Black T (EBT) clarifies the reaction mechanism dependency. The synergistic effect after azo compound addition is confirmed, and the time decreases about 23 and 13 min for RH and MB, respectively.

The photodecomposition acceleration effect is positively proportional to the azo dye concentration and no longer changes after reaching a specific equilibrium value. The comparison with the k values indicates higher rates for EBT > OG > MO > MR; the sequence is in agreement with the polarity of the four azo dye compounds. The azo compounds in the experiments were acid orange 7 (AO7), Congo red (CR), and amido black 10B (AB10B). The results were the same obtained for the other azo compounds confirming the synergistic oxidation effect.

The possible decomposition mechanism includes the highest occupied molecular orbital (HOMO) and the lowest unoccupied molecular orbital (LUMO) as a type of molecular frontier orbitals. Roughly, the HOMO level is for organic semiconductors, the equivalent of the valence band for the inorganic semiconductors, and the LUMO is the equivalent of the semiconductors' conduction band. The energy difference between them is called a HOMO-LUMO gap. The energy gap between the two frontier orbitals can be used to predict the strength and stability of the transition metal complexes and also their colors in solution.

The simulated changes of the azo molecule methylene orange (MO) in molecular energy structure in the photocatalyst Ag_2O surface indicate the LUMO composed by the atomic orbital contributions of benzene, nitrogen, and the nitrogen double bond, and the HOMO is a result of the paired electron orbits of negatively charged oxygen atoms in the sulfonic group.

After the MO adsorption in Ag_2O surface, the electrons on the surface of the Ag_2O can transfer to the molecules' LUMO and participate in feedback coordination at the bond. The C—N bond links the benzene ring to the azo group, and it becomes longer on adsorption. Such coordination effects weaken the π bonding conjugated over the whole molecular skeleton and start to be attacked by photogenerated electrons or radicals, with the presence of active fragmented intermediates. The resultant intermediates caused subsequent acceleration of azo bond cleavage also for non-azo organic cleavages due to their oxidative activity.

The infrared results of the MO adsorbed in Ag_2O indicate a shift for N═N bond from 1392 to 1396 cm^{-1}, for $-SO3^-$ was observed from 1314 and 1121 cm^{-1} to 1320 and 1120 cm^{-1}, and for C—N bonds from 817 and 749 cm^{-1} to 819 and 751 cm^{-1}. Another shift for Ag_2O was from 650 to 705 cm^{-1}. They are the confirmation of the sulfonic group acting as an electron donor and Ag_2O as an electron acceptor, weakening the conjugated π bonding and activating the N═N and C—N bonds confirming the adsorption of MO by Ag_2O. The MO peaks disappear including the sulfonic group and the azo bond after 2 min of irradiation, suggesting to be the first degraded group. The suspension started to be colorless, and the peaks assigned to be C═C bond start to appear after 5 min of exposition, an indication of the benzene rings broken. There is no observation of MO chemical structure after 10 min of light irradiation; it is an indication of the complete dye decomposition. The azo bond or C—N bond connected with the benzene rings broken first and produces active intermediates which accelerate the degradation of non-azo organics followed by the benzene ring broken.

Generally, the C—N bond linked to the benzene ring and the azo group of the MO is the first target for the free radicals produced by the photocatalyst. The possible intermediates are aminobenzenesulfonates, aromatic amines, phenolic compounds, and organic acids. The final products, also called mineralization, are N_2, CO_2, H_2O, SO_4^-, NH_4^-, and NO_3^-.

The presence of photodecomposition intermediates to the photodegradation process of RH and Ag_2O resulted in faster decomposition. The phenol addition reduces the decomposition time to 8 min, the sulfanilic acid in 10 min, and the benzoquinone in 30 min. However, the addition of acetic acid and hydroquinone slows the RH photodegradation. Finally, the MO molecule can be decomposed into holes or radicals generated over Ag_2O and break the C—N bonds releasing the benzene containing the intermediates such as benzene sulfonate and N-N dimethylaniline. The next intermediate products were hydroxybenzenesulfonate activated by excited Ag_2O and diffused in solution accelerating the degradation of organic compounds in the Ag_2O surface.

The description of the acceleration of the photodegradation process with azo dyes' presence in a mixed dye solution is a synergy between the azo structure and Ag_2O with the generation of aniline, sulfanilic acid, and phenol compounds which also accelerates the degradation of the non-azo compounds. The synergetic effect is beneficial for the Ag_2O photodecomposition applicability to treat the ordinary real wastewater with a complex dye mixture.

11. Conclusion

Environmental sustainability demands the advance in water treatment and the use of lighting natural resources. Brazil has one of the most stable and intense solar irradiation in the word. It has to be used not only for energy generation purposes but also and mostly for water treatment, water quality polishment, and furthermore water disinfection. The chapter performs a comparison of different green technologies for water treatment as natural solar irradiation. The photocatalytic hydroxide radicals are the photodecomposition potent oxidants and react fast and unselectively with surrounding chemical species via radical addition, hydrogen abstraction, or e^- transfer mechanisms. The transformation by-products of pharmaceuticals and EDC compounds (TBPs) with higher photodecomposition effectiveness ends up in complete mineralization with the production of CO_2, H_2O, and inorganic salts. The heterogeneous photocatalysis shows a strong dependence of the operating temperature, and the kinetics is usually dependent on the first step of the adsorption and the equilibrium modeled by Langmuir isotherms and Langmuir-Hinshelwood model. The first pseudo-order usually appears at the beginning of the reaction, just in the initial steps, and as the reaction proceeds, the intermediates' production could interfere with the radiation incidence. There is a competition of the adsorption sites of the catalyzer surface between the pollutant and others adsorbed species; the pollutants start concentration is a limiting reactant step with mass transfer limitations in lower concentrations. The semiconductor TiO_2 photocatalytic process has shown great potential as a low cost, environment-friendly treatment technology in degrading a wide range of pollutants with the formation of reactive oxygen species upon excitation of a semiconductor particle with light energy greater than the respective bandgap energy of the photocatalyst. The photocatalyst TiO_2 has superior characteristics over others with wide bandgap energy which requires the UV light which is 3–5% of natural solar light. The application of a variety of strategies improved the photocatalytic efficiencies from photocatalysts as dispersed solids to second-generation photocatalysts (chemically doped and physically modified dispersed solids) achieving better spectral sensitivity and photoactivity. Many studies indicate the scavengers' presence reduces the photodecomposition effect in water suspension. The ions HCO_3^-/CO_3^{2-}, SO_4^{2-}, Cl^- and NO_3^- showed inhibitory effects toward the hydroxyl radicals generated by AOPs; the natural organic matter (NOM) presence showed a synergistic effect increasing the E2 degradation, such degradation produces other radicals. The application of the visible light photodecomposition in a dye mixture of methylene blue (MB), methyl orange (MO), and rhodamine (RH) indicates the MO as the more stable azo compound than the other organic pollutants due to the aromatic groups attached at the end of the azo bond. Despite this fact when the light-driven photodecomposition uses the Ag_2O as a catalyzer, it was the fastest and easiest decomposed compound. Published results indicate the visible light photodecomposition with Ag_2O with the elimination of 90.2% of MO, 96.5% of RH, and 99.5% of MB using 4, 50, and 20 min, respectively. The photodecomposition acceleration synergistic effect is positively proportional to the azo dye concentration and no longer changes after reaching a specific equilibrium value. The comparison with the k values indicates higher rates for EBT > OG > MO > MR; the sequence is in agreement with the polarity of the four azo dye compounds. The azo compounds in the experiments were acid orange 7 (AO7), Congo red (CR), and amido black 10B (AB10B). The results were the same obtained for the other azo compounds confirming the synergistic oxidation effect. The possible decomposition mechanism includes the highest occupied molecular orbital (HOMO) and the lowest unoccupied molecular orbital (LUMO) as a type of molecular frontier orbitals. The description of the acceleration of the photodegradation process with azo dyes' presence in a mixed dye solution is a synergy between the azo structure and Ag_2O with the generation of aniline, sulfanilic acid, and phenol compounds which also accelerates the degradation of the non-azo compounds. The synergetic effect is beneficial for the Ag_2O photodecomposition applicability to treat the ordinary real wastewater with a complex dye mixture.

Acknowledgements

The author knowledge the National Council for Scientific and Technological Development (CNPq) and the Sao Paulo Research Foundation (Fapesp).

Author details

Nilce Ortiz
Institute for Nuclear and Energy Research–IPEN, São Paulo, Brazil

*Address all correspondence to: nortizbr@gmail.com

References

[1] Foteinis S, Borthwick AGL, Frontistis Z, Mantzavinos D. Environmental sustainability for light-driven processes for wastewater treatment applications. Journal of Cleaner Production. 2018;**82**:8-15. DOI: 10.1016/j.clepro.2018.02.038

[2] Kovacic M, Kopcic N, Kusic H, Bozic AL. Solar driven degradation of 17β estradiol using composite photocatalytic materials and artificial irradiation source: Influence of process and water matrix parameters. Journal of Photochemistry and Photobiology A: Chemistry. 2018;**361**:48-61. DOI: 10.1016/j.jphotochem.2018.05.015

[3] Quesada-Carera R, Mills A, Rourke CO. Action spectra of P25 TiO_2 and visible light adsorbing, carbon-modified titania in the photocatalytic degradation of stearic acid. Applied Catalysis B: Environmental. 2014;**150-151**:338-344. DOI: 10.1016/j.apcatb.2013.12.008

[4] Kamimura S, Miyazaki T, Zhang M, Li Y, Tsubota T, Ohno T. Au@Ag@Au double shell nanoparticles loaded on rutile TiO_2 for photocatalytic decomposition of 2-propanol under visible light irradiation. Applied Catalysis B: Environmental. 2016;**180**:255-262. DOI: 10.1016/j.apcatb.2017.04.028

[5] Haroune L, Salaun M, Menard A, Legault CY, Bellenger J. Photocatalytic degradation of carbamazepine and three derivatives using TiO_2 and ZnO; effect of pH, ionic strength, and natural organic matter. Science of the Total Environment. 2014;**475**:16-22. DOI: 10.1016/j.scitotenv2013.12.104

[6] Gao X, Peng W, Tang G, Guo Q, Luo Y. Highly efficient and visible—Light driven BiOCl for photocatalytic degradation of carbamazepine. Journal of Alloys and Compounds. 2018;**757**:455-465. DOI: 10.1016/j.jallcom.2018.05.081

[7] Canle M, Pérez MIF, Santaballa JA. Photocatalyzed degradation/abatement of endocrine disruptors. Current Opinion in Green and Sustainable Chemistry. 2017;**6**:101-138. DOI: 10.1016/j.cogsc.2017.06.008

[8] Gomes-Aviles A, Penas-Garzon M, Bedia J, Rodriguez JJ, Belver C. C-modified TiO_2 using lignin as carbon precursor for the solar photocatalytic degradation of acetaminophen. Chemical Engineering Journal. 2018;**358**:1574-1582. DOI: 10.1016/j.cej.2018.10.154

[9] Tobajas M, Belver C, Rodriguez JJ. Degradation of emerging pollutants in water under solar radiation using novel TiO_2-ZnO/clay nanoarchitectures. Chemical Engineering Journal. 2017;**309**:596-606. DOI: 10.1016/j.cej.2016.10.002

[10] Reddy PVL, Kim K, Kavitha B, Kumar V, Raza N, Kalagara S. Photocatalytic degradation of bisphenol A in aqueous media: A review. Journal of Environmental Management. 2018;**213**:189-205. DOI: 10.1016/j.envman.2018.02.059

[11] Kuo C, Wu C, Lin H. Photocatalytic degradation of bisphenol A in a visible light/TiO_2 system. Desalination. 2010;**256**:37-42. DOI: 10.1016/j.desal.2010.02.020

[12] Tsai W, Lee M, Su T, Chang Y. Photodegradation of bisphenol A in a batch TiO_2 suspension reactor. Journal of Hazardous Materials. 2009;**168**:269-275. DOI: 10.1016/j.hazmat.2009.02.034

[13] Bechambi O, Jlaiel L, Najjar W, Sayadi S. Photocatalytic degradation of bisphenol A in the presence of Ce-ZnO: Evolution of kinetics, toxicity and photodegradation mechanism. Materials Chemistry and Physics.

2016;**173**:95-105. DOI: 10.1016/j.matchemphys.2016.01.044

[14] Uyguner-Demirel CS, Birben NC, Bekbolet M. Elucidation of background organic matter matrix effect on the photocatalytic treatment of contaminants sing TiO_2: A review. Catalysis Today. 2017;**284**:202-214. DOI: 10.1016/j.cattod.2016.12.030

[15] Mboula VM, Héquet V, Andrés Y, Gru Y, Collin R, Dona-Rodriguez JM, et al. Photocatalytic degradation of estradiol under simulated solar light and assessment of estrogenic activity. Applied Catalysis B: Environmental. 2015;**162**:437-444. DOI: 10.1016/j.apcatb.2014.05.026

[16] Sun W, Li S, Mai J, Ni J. Initial photocatalytic degradation intermediates/pathways of 17 α ethynylestradiol: Effect of pH and methanol. Chemosphere. 2010;**81**:92-99. DOI: 10.1016/j.chemosphere.2010.06.051

[17] Serna-Galvis EA, Silva-Agredo J, Giraldo AL, Florez OA, Torres-Palma RA. Comparison of route, mechanism, and extent of treatment for the degradation of a β lactam antibiotic by TiO_2 photocatalysis, sonochemistry, electrochemistry, and the photo-Fenton system. Chemical Engineering Journal. 2016;**284**:953-962. DOI: 10.1016/j.cej.2015.08.154

[18] Serna-Galvis EA, Silva-Agredo J, Giraldo-Aguirre AL, Florez-Acosta OA, Torres-Palma RA. High-frequency ultrasound as an elective advanced oxidation process to remove penicillin antibiotics and eliminate its antimicrobial activity from the water. Ultrasonics Sonochemistry. 2016;**31**: 276-283. DOI: 10.1016/.ultsonch.2016.01.007

[19] Giraldo-Aguirre AL, Erazo-Erazo E, Florez-Acosta OA, Serna-Galvis EA, Torres-Palma R. $ATiO_2$ photocatalysis applied to the degradation and antimicrobial activity removal of oxacillin: Evaluation of matrix components, experimental parameters, degradation pathways and identification of organics by-products. Journal of Photochemistry and Photobiology A: Chemistry. 2015;**311**:95-103. DOI: 10.10167/j.photochem.2015.06.021

[20] Brigante M, Schulz PC. Remotion of the antibiotic tetracycline by titania and titania-silica composed materials. Journal of Hazardous Materials. 2011;**192**:197-1608. DOI: 10.1016/j.jhazmat.2011.06.082

[21] Bi N, Zheng H, Zhu Y, Jiang W, Liang B. Visible-light-driven photocatalytic degradation of non-azo dyes over Ag_2O and its acceleration by the addition of azo dye. Journal of Environmental Chemical Engineering. 2018;**6**:3150-3160. DOI: 10.1016/j.jece.2018.04.047

Chapter 6

Global Water Vapor Estimates from Measurements from Active GPS RO Sensors and Passive Infrared and Microwave Sounders

Shu-peng Ho and Liang Peng

Abstract

Water vapor plays an important role in both climate change processes and atmospheric chemistry and photochemistry. Global water vapor vertical profile can be derived from satellite infrared and microwave sounders. However, no single remote sensing technique is capable of completely fulfilling the needs for numerical weather prediction, chemistry, and climate studies in terms of vertical resolution, spatial and temporal coverage, and accuracy. In addition to the passive infrared and microwave sounder observations, the active global positioning system (GPS) radio occultation (RO) technique can also provide all-weather temperature and moisture profiles. In this chapter, we describe the current developments of global water vapor vertical profile and total precipitable water derived from active GPS RO measurements. In addition, we also demonstrate the potential improvement of global water vapor estimates using combined active GPS RO and passive IR/MW particularly from Atmospheric InfraRed Sounder (AIRS) and Advanced Technology Microwave Sounder (ATMS) measurements. Results show that because RO data are very sensitive to water vapor variation in the moisture rich troposphere, the RO data are able to provide extra water vapor information for the combined AIRS/ATMS and RO retrievals in the lower troposphere.

Keywords: water vapor, UT/LS, troposphere, greenhouse gas, climate process

1. Introduction

Water vapor is the major greenhouse gas in the atmosphere, which plays an important role in almost all the climate change processes. The transports and phase changes of water vapor directly affect the formation of cloud and precipitation, which modulate the hydrological cycle and the energy balance of the earth. Water vapor also has a strong effect on atmospheric chemistry and photochemistry. An accurate knowledge of the distribution of atmospheric water vapor is needed for climate change assessment, weather prediction, and atmospheric chemistry studies [1–3].

Global water vapor vertical profile can be derived from satellite infrared and microwave sounders (i.e., [4–8]). Nevertheless, no single remote sensing technique is able to completely fulfill the needs for numerical weather prediction, chemistry, and climate studies in terms of vertical resolution, spatial and temporal coverage, and accuracy. Satellite infrared (IR) and microwave sounders have been routinely used for monitoring the temperature and moisture profiles in the mid and lower troposphere since 1980. Launched in 2002, Atmospheric InfraRed Sounder (AIRS) is a high spectral resolution radiometer onboard NASA Aqua satellite. With those more than 2000 high spectrum resolution channels in infrared wavebands, AIRS is able to provide excellent temperature and water vapor retrievals at the mid-troposphere level under clear skies. The AIRS measurements, together with more recent high spectrum resolution infrared (IR) measurements from Atmospheric Sounding Interferometer (IASI, 2006–current), and Cross-track Infrared Sounder (CrIS, 2011–current) have maintained continuous observations of tropospheric water vapor since 2002. However, due to the limitation of resolving power in terms of weighting functions and signal to noise ratio of these instruments, accurate estimates of water vapor concentration in the lower troposphere (LT) are still not available. Infrared sounders cannot sense atmospheric profiles below clouds.

While infrared data are limited to clear skies, microwave (MV) sounders can provide all sky data products. There are three main microwave radiometers with sufficient resolution and stability to measure tropospheric water vapor: the Advanced Microwave Sounding Unit-A (AMSU-A) and Advanced Microwave Sounding Unit-B (AMSU-B) on NOAA-15, 16, and 17 (i.e., N15, N16, and N17) satellites, the Microwave Humidity Sounder (MHS) on NOAA-18 (N18) and MetOp-A (Meteorological Operational satellite A) satellites, and Advanced Technology Microwave Sounder (ATMS) on Suomi National Polar orbiting Partnership (SNPP) and the first Joint Polar Satellite System (JPSS-1). These microwave sounders onboard the polar-orbiting satellites have been routinely used by NOAA to generate the tropospheric temperature and moisture profiles for all-sky conditions and hydrological variables (e.g., rainfall, precipitable water, cloud water, ice water, etc.) under cloudy conditions.

Using a one-dimensional variational (1DVAR) scheme, the NOAA Microwave Integrated Retrieval System (MiRS) inversion package is routinely applied to the AMSU/MHS sensors onboard the NOAA-18 (N18), NOAA-19 (N19), and Meteorological Operational Satellite-A (Metop-A) satellites to optimally retrieve temperature, moisture, and surface skin temperature (SST), as well as hydrometer variables within clouds over land and oceans [9, 10]. The MiRS-retrieved parameters have been validated globally using independent measurements [9, 11, 12]. However, studies demonstrated that the MiRS-derived hydrometer parameters within clouds over lands and oceans still contain uncertainty, especially in the lower troposphere [10]. This is partly because there is not enough information from the AMSU/ATMS measurements to completely resolve the hydrometer variables, temperature, and water vapor profiles under clouds.

In addition to the passive infrared and microwave sounder observations, the active global positioning system (GPS) radio occultation (RO) technique can also provide all-weather temperature and moisture profiles. Unlike passive MW and IR sensors, GPS RO is an active remote sensing technique that can provide all-weather, high vertical resolution (from ~100 m near the surface to ~1.5 km at 40 km) bending angle, and refractivity profiles [13, 14]. With knowledge of the precise positions and velocities of the GPS and low earth orbiting (LEO) satellites, which carry the GPS receivers, a vertical distribution of bending angles at the ray perigee point (the point of the ray path that is closest to Earth) can be derived. From the vertical distribution of the bending angle, we can derive a vertical distribution of the atmospheric bending angle and refractivity, which is a function of atmospheric temperature, moisture, and pressure [13, 14]. The GPS RO data have been intensively used for weather prediction and climate studies since the launch of the

FORMOSAT-3/COSMIC (Formosa Satellite Mission #3/Constellation Observing System for Meteorology, Ionosphere & Climate) satellite in 2006 [13, 15–22]. Different numerical methods and various assumptions for bending angle calculation can be found in [17, 22].

In this chapter, we describe the current developments of global water vapor vertical profile and total precipitable water derived from active GPS RO measurements. We will also demonstrate the potential improvement of global water vapor estimates using combined active GPS RO and passive IR/MW measurements. We introduce the COSMIC temperature and moisture products in Section 2. The combined inversion algorithm is summarized in Section 3. The inversion simulation results are summarized in Section 4. The validation of inversion results from the combined RO-AMSU retrievals using collocated radiosonde observation (RAOBs) is shown in Section 5. We conclude this study in Section 6.

2. COSMIC temperature and water vapor retrievals

COSMIC observations distribute relatively uniformly in space and time (see **Figure 1**). By placing a GPS receiver in LEO, GPS RO technique measures the phase delay of the radio signal from the GPS constellation precisely as the signal traverses the Earth's atmosphere. Being an active limb-sounding measurement, GPS RO technique is capable of retrieving profiles of microwave refractivity at very high vertical resolution [23]. The root mean square (RMS) error was estimated to be less than 1 K based on a detailed theoretical study [23], and this estimate was found to be consistent with numerous cross-validation studies between RO, radiosonde observation (RAOB), and other satellite measurements (e.g., [20, 21, 24–28]).

Although RO measurements are not sensitive to clouds, they are very sensitive to the vertical structure of atmospheric density profiles (a function of temperature, pressure, and water vapor profiles). When accurate RO observations, precise positions, and velocities of GPS and LEO satellites are known, accurate atmospheric temperature and moisture profiles can be derived [15, 29]. In a neutral atmosphere,

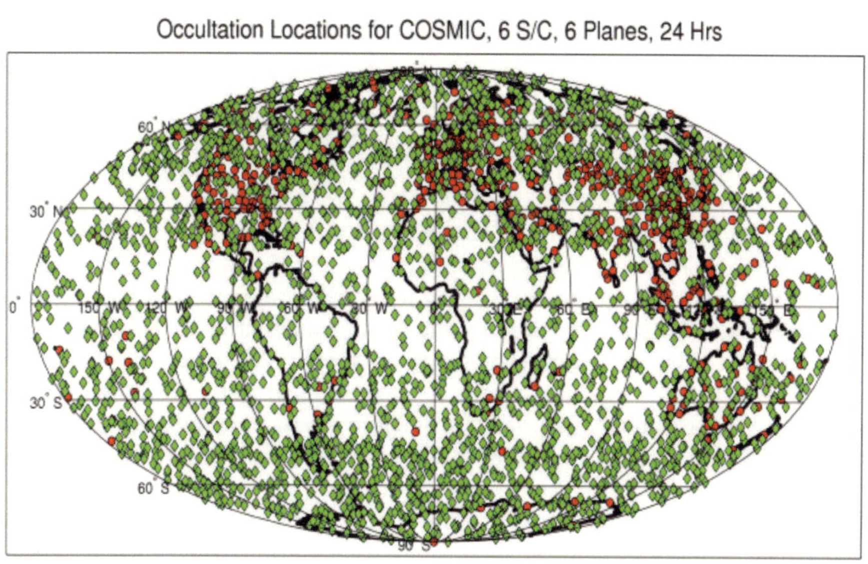

Figure 1.
Typical distribution of COSMIC GPS radio occultation soundings (green dots) over a 24-h period over the global. Red dots are the distribution of operational radiosonde stations.

the refractivity (N) is related to the pressure (P), the temperature (T), and the partial pressure of WV (PW) as represented by the following equation:

$$N = 77.6\frac{P}{T} + 3.73 \times 10^5 \frac{P_W}{T^2} \quad (1)$$

Above the UT (~8 km) where moisture is negligible, the dry temperature and the actual temperature are nearly equal [23]. The accuracy, precision, and long-term stability of RO data have been quantified by many studies under various atmospheric conditions [18–21, 26–28, 30]. GPS RO measurements have many important attributes that make them uniquely suitable for climate monitoring. These include: (i) no satellite-to-satellite bias, (ii) no instrument drift, and (iii) not affected by clouds and precipitation.

2.1 COSMIC temperature uncertainty

The distribution of water vapor profile depends on temperature [31, 32]. COSMIC Data Analysis and Archive Center (CDAAC) is an international operational center that routinely inverses RO excess phase data obtained from multiple RO mission to bending angle and refractivity profiles. Recently, CDAAC has developed and installed new and improved RO inversion software, including improvements to precise orbit determination (POD), excess atmospheric phase computation, and neutral atmospheric inversion processing. Now these consistent RO inversion algorithms are applied to several international RO missions to derive the vertical distribution of bending angle, refractivity, temperature, and geo-potential height. These RO missions include GPS/MET (from 1995 to 1997, no overlap with other RO missions), COSMIC (launched in April 2006), Challenging Mini-satellite Payload (CHAMP, from 2001 to 2008), Gravity Recovery And Climate Experiment (GRACE, launched in 2004), Satélite de Aplicaciones Científicas-C (SAC-C, launched in 2000), GNSS RO Receiver for Atmospheric Sounding (GRAS, launched in 2007), Communication/Navigation Outage Forecast System (C/NOFS, launched in 2008), and Terra Synthetic Aperture Radar (SAR) operating in the X-band (TerraSAR-X, launched in 2007).

Currently, multi-year GPS RO climate data can be obtained from the GeoForschungsZentrum Potsdam (GFZ), Germany, the Jet Propulsion Laboratory (JPL), Pasadena, CA, USA, the University Corporation for Atmospheric Research (UCAR), Boulder, CO, USA, and the Wegener Center of the University of Graz (WegC), Graz, Austria. Different centers have used different assumptions, initializations, and implementations in the excess phase processing and inversion procedures, which may introduce refractivity differences between centers. Ho et al. [17] have used 5 years (2002–2006) of monthly mean climatologies (MMC) of retrieved refractivity from CHAMP generated by the above four centers to quantify the processing procedure-dependent errors. Results show that the absolute values of fractional refractivity anomalies among the centers are in general ≤0.2% from 8 to 25 km altitude (not shown). The median absolute deviation among the centers is less than 0.2% globally. This provides useful bounds on the errors introduced by data processing schemes.

The near real-time, postprocessing, and reprocessing status for all RO missions at CDAAC are summarized on the CDAAC website (http://cdaac-www.cosmic.ucar.edu/cdaac/products.html). A new version of Metop-A reprocessing, named Metop-A2016, has just finished processing and validation and will be posted to the CDAAC website in the next month. To illustrate the consistency between COSMIC and Metop-A CDAAC products, **Figure 2** shows a statistical comparison for July 2014 of co-located COSMIC postprocessed (consistent with COSMIC2013) and Metop-A2016 reprocessed bending angle profiles. The match criteria used were: time differences <90 min and distances <250 km. The mean differences are very small. The differences in standard deviation at ~30 km altitude (south pole region larger than north pole region) are believed to be due to stronger horizontal variability of the atmosphere during local winter. Remaining GPS RO missions will be processed as soon as possible, and all data will be made available via CDAAC's website.

RO derived temperature profiles especially in the lower stratosphere have also been intensively validated. **Figure 3** depicts that COSMIC temperature is very close to those from Vaisala-RS92 from 200 to 20 hPa (around 12 to 25 km). Note that, Vaisala-RS92 is one of the most accurate modern radiosondes where the structural uncertainties are ±0.2 K below 100 hPa and somewhat higher at higher levels. According to RS92 data continuity link under the Vaisala website, the Vaisala data including RS92 have been corrected for possible radiation errors. Their mean temperature difference in this height range is very close to zero. Because the quality of RO data does not vary with geophysical location and time, it is very useful to assess systematic errors of radiosonde sensors, whose characteristics may be affected by the changing environment and sensor types (e.g., [26, 28]). This result also shows the quality of RO temperature profiles where the precision of the mean of COSMIC-derived temperature profiles is estimated to be better than 0.05 K from 8 to 30 km [18].

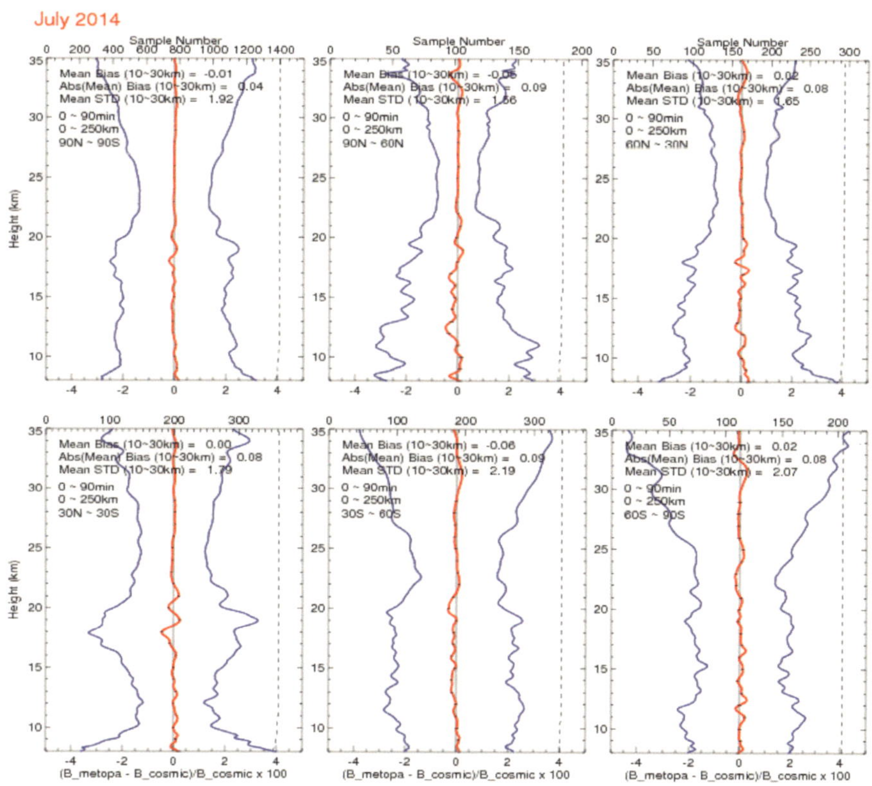

Figure 2.
Statistical comparison of co-located COSMIC postprocessed and Metop-A reprocessed bending angle profiles (red = mean and blue = standard deviation). Match criteria: time differences <90 min and distances <250 km.

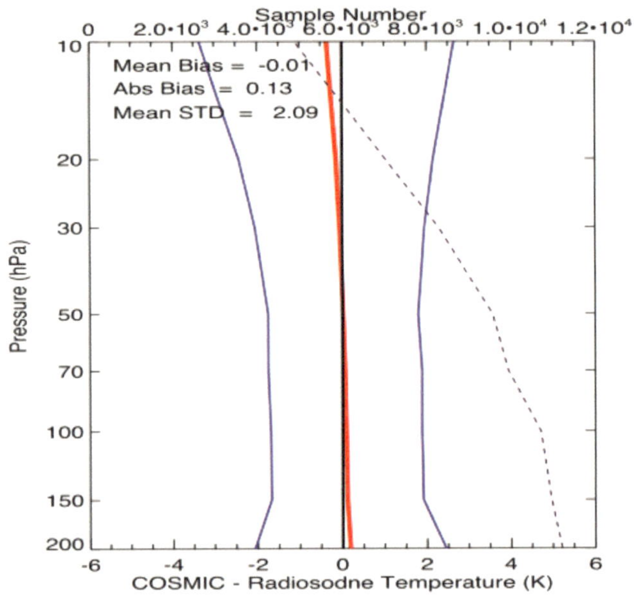

Figure 3.
Comparisons of temperature between COSMIC and radiosonde for Vaisala-RS92. Mean bias, absolute mean bias, and mean standard deviation from 200 to 10 hPa are computed. The red line is the mean difference, the blue line is the standard deviation, and the dotted line is the sample number. The top X axis shows the sample number.

2.2 COSMIC water vapor uncertainty

By using an advanced tracking technique, known as "open-loop tracking" [33], more than 90% of RO profiles from the COSMIC mission penetrate to below 2 km. As shown in Eq. (1), GPS RO refractivity is sensitive to temperature or water vapor, depending on the atmospheric conditions. Ho et al. [34] showed that in the upper troposphere where water vapor is negligible, RO observations are more sensitive to atmospheric temperature variations than to water vapor content. However, in the moisture rich troposphere, the RO refractivity is more sensitive to water vapor variation [34].

A 1D-var algorithm (http://cosmic-io.cosmic.ucar.edu/cdaac/doc/documents/1d var.pdf) is used to derive optimal temperature and water vapor profiles while temperatures and water vapor profiles from RO refractivity (see Eq. (1)). The ERA-Interim reanalyzes are used as a priori estimates for the 1D-var algorithm. The accuracy of COSMIC-derived total precipitation water (TPW) has been demonstrated by comparisons with TPW derived from ground-based GPS (i.e., International Global Navigation Satellite Systems (IGS, [20, 35])) which are assumed not to be geolocation dependent. **Figure 4** (left panel) depicts COSMIC TPW and those from ground-based GPS collected within 2 h and 200 km in 2008. Only those COSMIC profiles whose lowest penetration heights are within 200 m of the height of ground-based GPS stations are included. Results demonstrate that the mean TPW difference between IGS and COSMIC is less than 0.2 mm with a standard deviation of 2.69 mm. This demonstrates the accuracy of COSMIC-derived water vapor in the lower troposphere, which should be particularly useful for improving AIRS/ATMS retrievals, especially over ice and cold surface backgrounds (see Sections 4 and 5). The right panel in **Figure 4** shows the time series of the COSMIC TPW and those from IGS at the same station. This demonstrates the importance and usefulness of COSMIC RO observations in depicting global water vapor variations.

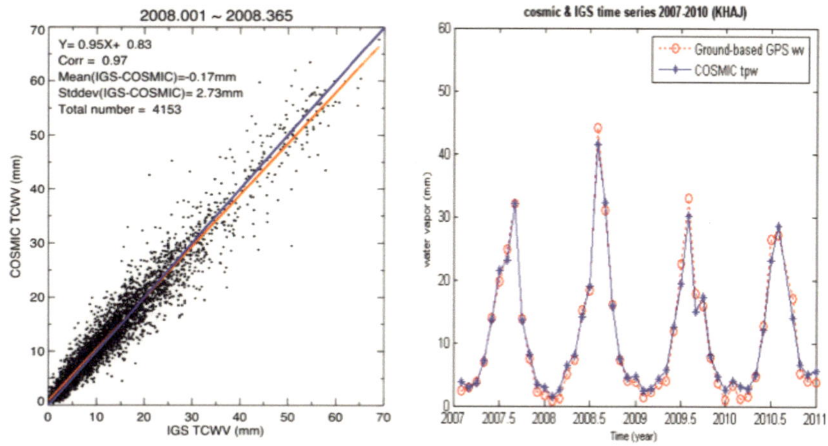

Figure 4.
Left panel: the global comparisons of total precipitable water (TPW) between COSMIC and those derived from ground-based GPS (i.e., IGS) for 2008. The right panel indicates the time series of IGS TPW and COSMIC-derived TPW near the IGS station.

COSMIC data have been used to study atmospheric temperature and refractivity trends in the lower stratosphere [17–19, 22] and variation of water vapor above, within, and below clouds [36–43]. COSMIC water vapor data have also been used to detect climate signals like El Niño-Southern Oscillation (ENSO; [38–40]), Madden-Julian Oscillation (MJO; [41]), atmospheric rivers [42, 43], and TPW variation owing to global warming [44–46].

In this study, we will use COSMIC data collocated with AIRS (COSMIC-AIRS pairs) and ATMS (COSMIC-ATMS pairs) to derive the temperature and moisture profiles. The relatively uniformly distributed COSMIC profiles in space and time would allow numerous RO and AIRS/ATMS coincident pairs, which would provide unprecedented atmospheric temperature and moisture profiles under various atmospheric conditions, which was not possible before.

3. The RO-ATMS and RO-AIRS inversion algorithms

3.1 RO-ATM inversion algorithm

As mentioned above, MiRS is a MW inversion package used by NOAA to perform a physical-based microwave retrieval in all-weather scenarios over all land-surface types. The MiRS data products have been routinely assessed using independent measurements [9, 11, 12]. The MiRS package can be downloaded, for free, at http://mirs.nesdis.noaa.gov/download.php. In this study, we revised the current MiRS algorithm and use measurements from ATMS collocate with RO data to develop an enhanced RO-ATMS inversion algorithm (i.e., RO-MiRS). We included the RO refractivity forward operator (Eq. (1)) into the RO-MiRS algorithm.

3.2 RO-AIRS inversion algorithm

Ref. [34] has detailed the information content for AIRS, RO, and the combined AIRS with RO for temperature and water vapor retrievals. Similar to the AIRS V6, the RO-AIRS inversion system is a 1D-var physical inversion system that retrieves the temperature and water vapor profiles sequentially. The updated fast and accurate AIRS transmittance model (Standard Alone AIRS-Radiance Transfer Algorithm package—SARTA [47]) is served as an AIRS forward operator. The configurations of background covariance matrix, error covariance matrix, and *a priori* used in the concurrent RO-AIRS retrieval are detailed in [48] and are not further described here.

4. Simulation inversion of the combined measurements in the troposphere and lower stratosphere

4.1 Temperature and water vapor profiles derived from the combined RO-AIRS and RO-MW measurements in the troposphere and the boundary layer

To illustrate how the collocated RO data benefit the AIRS retrievals, we conduct a multiple variable regression simulation study to inverse RO and AIRS measurements simultaneously to obtain the temperature and water vapor profiles. The SARTA [47] is used to simulate the AIRS radiances. The simulated AIRS brightness temperatures (BTs) and RO refractivity measurements are computed by applying the NOAA-88b temperature and moisture profiles to AIRS and RO refractivity forward operators plus the known AIRS instrument noises and RO refractivity measurement noises, respectively.

The temperature and moisture root mean square errors (RMSEs) for RO, AIRS, and the combined RO and AIRS retrievals are plotted in **Figure 5**. It is demonstrated in **Figure 5** that the combined AIRS and RO observations act to constrain the individual solutions. The significantly improved water vapor RMSE is found in both the middle and lower troposphere. The RMSEs of water vapor mixing ratio for AIRS and RO improved from 1.5 and 1.0 g/kg at surface, respectively, to 0.5 g/kg for the GPS RO combined AIRS retrievals. Since GPS refractivity is less sensitive to temperature in the troposphere, only small temperature RMSE improvements are found.

The synergy of using RO observations with microwave observations has been demonstrated in numerous studies, including the comparison of Microwave Sounding Unit (MSU)/AMSU climate records with RO data in the upper troposphere [19], evaluating the accuracy of Special Sensor Microwave Imager (SSMI) water vapor retrievals [44–46, 49], and developing methods for concurrent

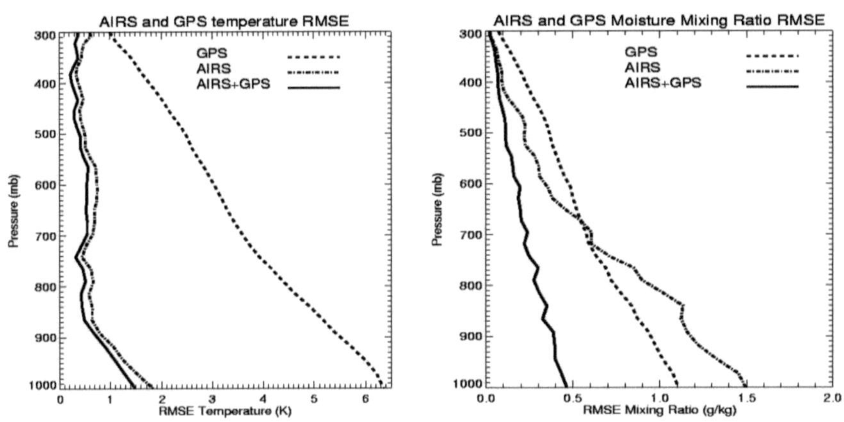

Figure 5.
The RMSE of retrieval results for temperature (left panel) and water vapor mixing ratio (right panel) for AIRS, GPS, and AIRS combined with GPS measurements.

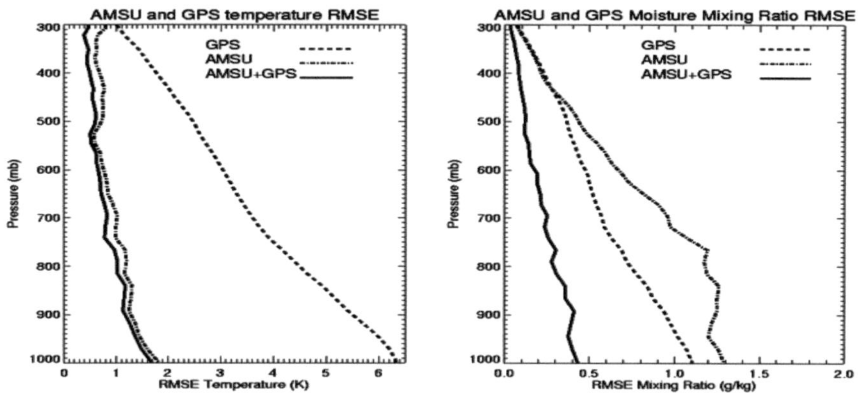

Figure 6.
The RMSE of retrieval results for temperature (left panel) and water vapor mixing ratio (right panel) for AMSU, GPS RO, and ASMU combined with GPS RO measurements over the globe.

inversion of RO and Atmospheric Infrared Sounder (AIRS) data [34, 48]. To illustrate how the increased amounts of RO data benefit the AMSU retrievals, we conduct a multiple variable regression simulation study using RO and MW measurements that are inversed simultaneously to retrieve the temperature and water vapor profiles. **Figure 6** shows the temperature and moisture RMSEs from the multiple variable regression simulation study of the GPS RO, AMSU, and the combined GPS RO and AMSU retrievals. The simulated AMSU BTs and RO refractivity measurements are computed by applying the NOAA-88b temperature and moisture profiles to AMSU and RO refractivity forward operators plus the known AMSU instrument noises and RO refractivity measurement noises, respectively.

It is shown in the right panel of **Figure 6** that because RO data are very sensitive to water vapor variation, the RMSEs for AMSU water vapor mixing ratio at the surface decreases from 1.3 g/kg (for AMSU only retrievals) to 0.4 g/kg when both GPS RO and AMSU data are used. The left panel of **Figure 6** shows that AMSU temperature measurements tremendously improve the GPS RO temperature retrieval when both GPS RO and AMSU data are used. These retrieval results demonstrate that the nadir viewing AMSU and limb-viewing GPS observations act to constrain the individual solutions; therefore providing much improved water vapor retrievals, particularly in the middle and lower troposphere. This is also demonstrated that by adding the RO refractivity operator in the AIRS inversion package described in [48], we are able to constrain the temperature and moisture profiles from the RO-AIRS observations and obtain the improved atmospheric thermal structure, which was not possible for individual sensors.

4.2 Improving the temperature and water vapor retrievals in the upper troposphere and stratosphere using combined AIRS, AMSU, and GPS RO measurements

In this section, we used GPS RO data to constrain the AIRS and AMSU temperature retrievals serving to improve moisture retrievals in the upper troposphere and lower stratosphere (UT/LS). In the upper troposphere, GPS RO refractivity is very sensitive to temperature but less sensitive to moisture. It is demonstrated by Ho et al. [34] that GPS RO refractivity can resolve temperatures greater than 1 K around 200 hPa but it can only sense about 15% of water vapor variation. **Figure 7** shows the temperature and moisture retrieval RMSE for AIRS, AMSU, and GPS RO as well as the combined AIRS, AMSU, and GPS RO data. The multi-variable

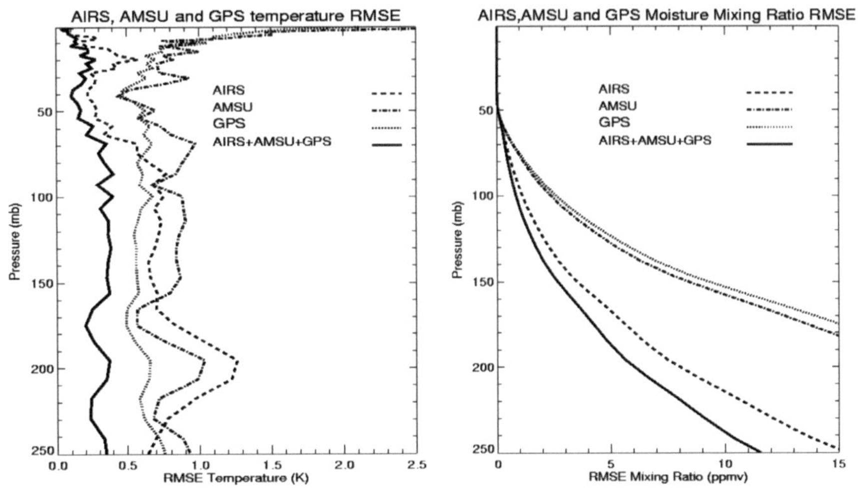

Figure 7.
The RMSE of temperature for AIRS, AMSU, GPS, and AIRS+AMSU+GPS in the UT/LS is on the left panel, and the RMSE of water vapor for AIRS, AMSU, GPS, and AIRS+AMSU+GPS in the UT/LS is on the right panel.

regression method described in Section 4.1 is used for the data from the simulation study. The 100 level AIRS vertical grids are used for all AIRS, AMSU, and GPS RO data. With a very high vertical resolution GPS RO refractivity profiles, AIRS and AMSU temperature RMSEs improved from 0.8 and 1.0 K, respectively, between 250 and 100 hPa (tropopause layer) to 0.4 K, which lead to AIRS and AMSU moisture RMSE around the same layer decrease from 4 and 15 ppmv, respectively, to around than 3 ppmv.

Since the open-loop tracking algorithm is only applied to COSMIC data, GPS RO data from COSMIC are used with AIRS data to derive moisture and temperature profiles in the clear skies of the free troposphere. For the UT/LS retrievals, GPS RO from SAC-C, GRACE, CHAMP, COSMIC, and GRAS data can be used to collocate with AIRS data. It can be seen in the left panel of **Figure 6** that the region of the largest temperature gradient is around 200 hPa, where the temperature RMSE for AIRS and AMSU is around 1.0 K. The fact that much improved temperature retrievals from GPS RO data (RMSE is ~0.6 K) and from the combined AIRS, AMSU, and GPS RO data (RMSE is ~0.3 K) are very useful to construct accurate temperature and moisture structures in the UT/LS region for the entire globe.

5. Initial results for the RO-MiRS retrievals from the COSMIC and ATMS data

We have successfully implemented the RO refractivity forward operator (Eq. (1)) into the current MiRS Version 11. This initial experiment is to demonstrate the feasibility of the proposed fusion approach to simultaneously retrieve global temperature and water profiles and hydrological data products using MiRS from the current operational COSMIC and ATMS data. Ten days of COSMIC-ATMS pairs are collected and inverted using RO-MiRS. **Figures 8** and **9** compared the co-located COSMIC and ATMS pairs (collected within 100 km and 15 min) with those temperature and moisture measurements from the Vaisala-RS92 radiosondes, respectively. **Figure 8** depicts that the mean temperature biases for RO/ATMS relative to

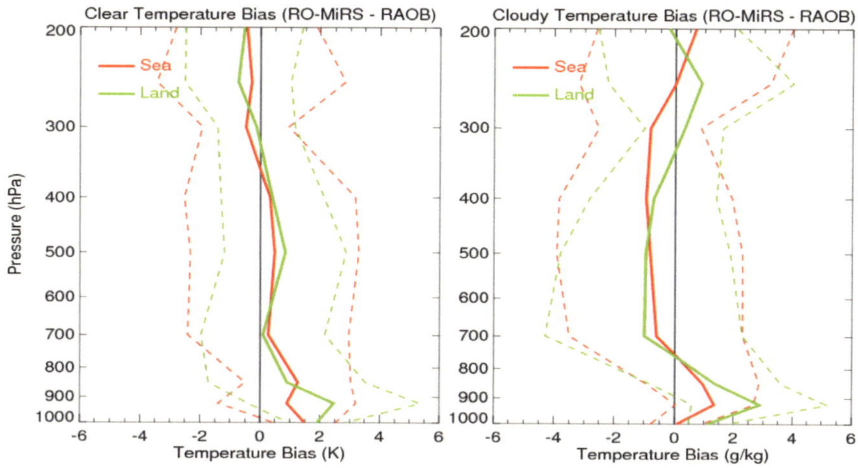

Figure 8.
RO-MiRS COSMIC/ATMS retrieved temperature bias with respect to co-located RS92 measurements (red: ocean and green: land) for clear (left) and rainy (right) conditions.

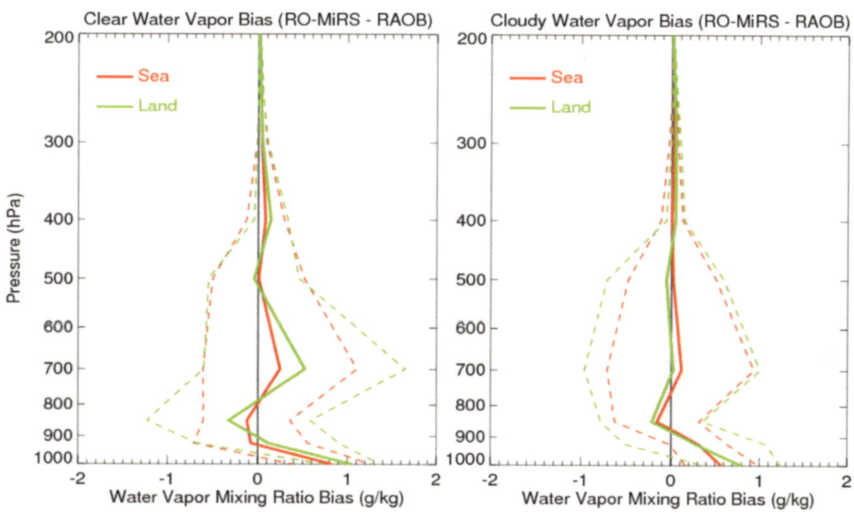

Figure 9.
RO-MiRS COSMIC/ATMS retrieved water vapor bias with respect to co-located RS92 measurements (red: ocean and green: land) for clear (left) and rainy (right) conditions.

those from RS92 are equal to 0.39 K with a standard deviation of 2.26 K for clear ocean cases and 0.0 K with a standard deviation of 2.35 K for cloudy ocean cases, respectively. The temperature biases for RO/ATMS results are larger from those over land than from those over oceans for both clear and cloudy conditions.

Figure 9 shows comparison results for RO/ATMS global water vapor retrievals. Over ocean surfaces, the retrieved bias is relatively low at all layers. The land retrievals show a larger bias than those over oceans. The global mean water vapor biases from surface to 200 mb for clear/sea, clear land, cloudy/sea, and cloudy/land cases are 0.11, 0.17, 0.11, and 0.11 g/kg, respectively. The temperature and water vapor biases for ATMS-only retrievals relative to those from RS92 are about 10–20% larger than the COSMIC/ATMS results at different levels (not shown).

6. Conclusions

In this study, we summarized research studies for quantifying global water vapor variation estimated using measurements from active GPS RO sensors and passive infrared and microwave sounders. A new inversion algorithm by inverting the GPS RO observations collocated with the NASA Aqua AIRS measurements to retrieve enhanced temperature and water vapor profiles is introduced. This effort is meant to generate improved temperature and moisture profiles, which are not possible by each individual sensors at the locations and times for the RO-AIRS collocated pairs. In addition, we also introduce a new approach for retrieving RO data with collocated MW measurements. By including the RO refractivity forward operator into the currently available MiRS package, we are able to provide the improved temperature and moisture profile in the troposphere.

In the combined RO and AIRS retrieval (for simulation experiments), the high vertical resolution RO retrieved temperature profiles are able to help to resolve the sharp temperature inversion layer in the UT/LS (i.e., the tropopause) and constrain AIRS water vapor retrieval in the same altitude. Because RO data are also very sensitive to water vapor variation in the moisture rich troposphere, the RO data shall also help to provide extra water vapor information for the combined AIRS and RO retrievals in the lower troposphere. It is demonstrated that the combined AIRS and RO observations act to constrain the individual solutions, the significantly improved water vapor RMSE is found in both the middle and lower troposphere. The RMSEs of water vapor mixing ratio for AIRS and GPS RO improved from 1.5 and 1.0 g/kg at surface, respectively, to 0.5 g/kg for the GPS RO combined AIRS retrievals. Since GPS refractivity is less sensitive to temperature in the troposphere, only small temperature RMSE improvements are found. Similar results are found in the COSMIC, ATMS, and ATMS+COSMIC retrieval results.

In future, we will apply AIRS and COSMIC data from 2006 to 2016 to the derived physical inversion algorithm and validate the retrieval results against in situ data. COSMIC's success has also prompted U.S. agencies to move forward with a follow-on FORMOSAT-7/COSMIC-2 (hereafter COSMIC-2) RO mission with Taiwan. The mission will launch six satellites into low-inclination orbits in early 2018 which is expected to yield up to 6000 uniformly distributed RO profiles per day. This would allow numerous RO and AMSU/ATMS coincident pairs after 2018, which would provide unprecedented atmospheric thermal and hydrometer information below clouds under various atmospheric conditions, which was not possible before.

Author details

Shu-peng Ho* and Liang Peng
COSMIC Project Office, University Corporation for Atmospheric Research, Boulder, Colorado, USA

*Address all correspondence to: spho@ucar.edu

References

[1] Hartmann DL. Climate change, tropical surprises. Science. 2002;**295**: 811-812

[2] Crook NA. Sensitivity of moist convection forced by boundary layer processes to low-level thermodynamic fields. Monthly Weather Review. 1996; **124**:1767-1785

[3] Mason BJ. Recent improvements in numerical weather prediction. GARP Publication Series 26. 1986;**II**: 295-311

[4] Wentz FJ, Spencer RW. SSM/I rain retrievals within a unified all-weather ocean algorithm. Journal of the Atmospheric Sciences. 1998;**56**: 1613-1627

[5] Fetzer EJ, Lambrigtsen BH, Eldering A, Aumann HH, Chahine MT. Biases in total precipitable water vapor climatologies from atmospheric infrared sounder and advanced microwave scanning radiometer. Journal of Geophysical Research. 2006;**111**: D09S16. DOI: 10.1029/2005JD006598

[6] John VO, Soden BJ. Temperature and humidity biases in global climate models and their impacts on climate feedbacks. Geophysical Research Letters. 2007;**34**:L18605. DOI: 10.1029/2007GL030736

[7] Fetzer EJ, Read WG, Waliser D, Kahn BH, Tian B, Vömel H, et al. Comparison of upper tropospheric water vapor observations from the microwave limb sounder and atmospheric infrared sounder. Journal of Geophysical Research. 2008;**113** (D22):D22110

[8] Noël S, Buchwitz M, Burrows JP. First retrieval of global water vapour column amounts from SCIAMACHY measurements. Atmospheric Chemistry and Physics. 2004;**4**:111-125

[9] Boukabara S-A, Garrett K, Chen W, Liu Q, Yan B, Weng F. Global coverage of total precipitable water using a microwave variational algorithm. IEEE Transactions on Geoscience and Remote Sensing. 2010;**48**(10):3608-3621

[10] Boukabara S-A, Garrett K, Chen W, Iturbide-Sanchez F, Grassotti C, Kongoli C, et al. MiRS: An all-weather 1DVAR satellite data assimilation & retrieval system. IEEE Transactions on Geoscience and Remote Sensing. 2011; **49**(9):3249-3272. DOI: 10.1109/TGRS.2011.2158438

[11] Iturbide-Sanchez F, Boukabara S-A, Chen R, Garrett K, Grassotti C, Chen W, et al. Assessment of a variational inversion system for rainfall rate over land and water surfaces. IEEE Transactions on Geoscience and Remote Sensing. 2011;**49**(9):3311-3333. DOI: 10.1109/TGRS.2011.2119375

[12] Kongoli C, Boukabara S-A, Yan B, Weng F, Ferraro R. A new sea-ice concentration algorithm based on microwave surface emissivities. IEEE Transactions on Geoscience and Remote Sensing. 2011;**49**(1):175-189

[13] Anthes RA, Bernhardt PA, Chen Y, Cucurull L, Dymond KF, Ector D, et al. The COSMIC/FORMOSAT-3 mission-early results. Bulletin of the American Meteorological Society. 2008;**89**:313-333

[14] Anthes RA. Exploring Earth's atmosphere with radio occultation: Contributions to weather, climate and space weather. Atmospheric Measurement Techniques. 2011;**4**: 1077-1103. DOI: 10.5194/amt-4-1077-2011

[15] Kuo Y-H, Wee T-K, Sokolovskiy S, Rocken C, Schreiner W, Hunt D, et al. Inversion and error estimation of GPS radio occultation data. Journal of the

Meteorological Society of Japan. 2004; **82**:507-531

[16] Foelsche U, Pirscher B, Borsche M, Kirchengast G, Wickert J. Assessing the climate monitoring utility of radio occultation data: From CHAMP to FORMOSAT-3/COSMIC. Terrestrial, Atmospheric and Oceanic Sciences. 2009;**20**:155-170

[17] Ho S-P, Kirchengast G, Leroy S, Wickert J, Mannucci AJ, Steiner AK, et al. Estimating the uncertainty of using GPS radio occultation data for climate monitoring: Inter-comparison of CHAMP refractivity climate records 2002–2006 from different data centers. Journal of Geophysical Research. 2009;**114**:D23107. DOI: 10.1029/2009JD011969

[18] Ho S-P, Goldberg M, Kuo Y-H, Zou C-Z, Schreiner W. Calibration of temperature in the lower stratosphere from microwave measurements using COSMIC radio occultation data: Preliminary results. Terrestrial, Atmospheric and Oceanic Sciences. 2009;**20**: 87-100. DOI: 10.3319/TAO.2007.12.06. 01(F3C)

[19] Ho S-P, He W, Kuo Y-H. Construction of consistent temperature records in the lower stratosphere using global positioning system radio occultation data and microwave sounding measurements. In: Steiner AK et al., editors. New Horizons in Occultation Research. Berlin: Springer; 2009. pp. 207-217. DOI: 10.1007/978-3-642-00321-9_17

[20] Ho S-P, Kuo Y-H, Schreiner W, Zhou X. Using SI-traceable global positioning system radio occultation measurements for climate monitoring [in "States of the Climate in 2009"]. Bulletin of the American Meteorological Society. 2010;**91**(7):S36-S37

[21] Ho S-P, Zhou X, Kuo Y-H, Hunt D, Wang J-H. Global evaluation of radiosonde water vapor systematic biases using GPS radio occultation from COSMIC and ECMWF analysis. Remote Sensing. 2010;**2**(5):1320-1330

[22] Ho S-P, Hunt D, Steiner AK, Mannucci AJ, Kirchengast G, Gleisner H, et al. Reproducibility of GPS radio occultation data for climate monitoring: Profile-to-profile inter-comparison of CHAMP climate records 2002 to 2008 from six data centers. Journal of Geophysical Research. 2012;**117**:D18111. DOI: 10.1029/2012JD017665

[23] Kursinski ER, Hajj GA, Schofield JT, Linfield RP. Observing Earth's atmosphere with radio occultation measurements using the global positioning system. Journal of Geophysical Research. 1997;**102**(D19): 23429-23465

[24] Rocken C et al. Analysis and validation of GPS/MET data in the neutral atmosphere. Journal of Geophysical Research. 1997;**102**(D25): 29849-29866

[25] Kuo Y-H, Schreiner WS, Wang J, Rossiter DL, Zhang Y. Comparison of GPS radio occultation soundings with radiosondes. Geophysical Research Letters. 2005;**32**:L05817. DOI: 10.1029/ 2004GL021443

[26] He W, Ho S, Chen H, Zhou X, Hunt D, Kuo Y. Assessment of radiosonde temperature measurements in the upper troposphere and lower stratosphere using COSMIC radio occultation data. Geophysical Research Letters. 2009;**36**:L17807. DOI: 10.1029/ 2009GL038712

[27] Ho S-P, Peng L, Mears C, Anthes R. Comparison of global observations and trends of total precipitable water derived from microwave radiometers and COSMIC radio occultation from 2006 to 2013. Atmospheric Chemistry and Physics. 2018;**18**:259-274. https:// doi.org/10.5194/acp-18-259-2018

[28] Ho S-P, Peng L, Voemel H. Characterization of the long-term radiosonde temperature biases in the upper troposphere and lower stratosphere using COSMIC and Metop-A/GRAS data from 2006 to 2014. Atmospheric Chemistry and Physics. 2017;**17**:4493-4511. DOI: 10.5194/acp-17-4493-2017

[29] Hajj GA, Ao CO, Iijima BA, Kuang D, Kursinski ER, Mannucci AJ, et al. CHAMP and SAC-C atmospheric occultation results and inter-comparisons. Journal of Geophysical Research. 2004; **109**:D06109. DOI: 10.1029/2003JD003909

[30] Ho S-P, Kuo YH, Zeng Z, Peterson T. A comparison of lower stratosphere temperature from microwave measurements with CHAMP GPS RO data. Geophysical Research Letters. 2007;**34**:L15701. DOI: 10.1029/2007GL030202

[31] Trenberth KE, Guillemot CJ. Evaluation of the atmospheric moisture and hydrological cycle in the NCEP/NCAR reanalyses. Climate Dynamics. 1998;**14**:213-231

[32] Trenberth KE, Fasullo J, Smith L. Trends and variability in column integrated atmospheric water vapor. Climate Dynamics. 2005;**24**:741-758

[33] Sokolovskiy S, Kuo Y-H, Rocken C, Schreiner WS, Hunt D, Anthes RA. Monitoring the atmospheric boundary layer by GPS radio occultation signals recorded in the open-loop mode. Geophysical Research Letters. 2007;**33**: L12813. DOI: 10.1029/2006GL025955

[34] Ho S-P, Kuo YH, Sokolovskiy S. Improvement of the temperature and moisture retrievals in the lower troposphere using AIRS and GPS radio occultation measurements. Journal of Atmospheric and Oceanic Technology. 2007;**24**:1726-1173. DOI: 0.1175/JTECH2071.1

[35] Wang J, Zhang LY, Dai A, Van Hove T, Van Baelen J. A near-global, 2-hourly data set of atmospheric precipitable water vapor from ground-based GPS measurements. Journal of Geophysical Research. 2007;**112**:D11107

[36] Biondi R, Randel W, Ho S-P, Neubert T, Syndergaard S. Thermal structure of intense convective clouds derived from GPS radio occultations. Atmospheric Chemistry and Physics. 2012;**12**:5309-5318. DOI: 10.5194/acp-12-5309-2012

[37] Biondi R, Ho S-P, Randel W, Neubert T, Syndergaard S. Tropical cyclone cloud-top heights and vertical temperature structure detection using GPS radio occultation measurements. Journal of Geophysical Research. 2013; **118**:1-13. DOI: 10.1002/jgrd.50448

[38] Teng W-H, Huang C-Y, Ho S-P, Kuo Y-H, Zhou X-J. Characteristics of global precipitable water in ENSO events revealed by COSMIC measurements. Journal of Geophysical Research. 2013;**118**:1-15. DOI: 10.1002/jgrd.50371

[39] Scherllin-Pirscher B, Deser C, Ho S-P, Chou C, Randel W, Kuo Y-W. The vertical and spatial structure of ENSO in the upper troposphere and lower stratosphere from GPS radio occultation measurements. Geophysical Research Letters. 2012;**39**:L20801. DOI: 10.1029/2012GL053071

[40] Huang C-Y, Teng W-H, Ho S-P, Kuo Y-H. Global variation of COSMIC precipitable water over land: Comparisons with ground-based GPS measurements and NCEP reanalyses. Geophysical Research Letters. 2013;**40**: 5327-5331. DOI: 10.1002/grl.50885

[41] Zeng Z, Ho S-P, Sokolovskiy S. The structure and evolution of Madden-Julian oscillation from FORMOSAT-3/COSMIC radio occultation data. Journal of Geophysical Research. 2012;**117**: D22108. DOI: 10.1029/2012JD017685

[42] Ma Z, Kuo Y-H, Ralph FM, Neiman PJ, Wick GA, Sukovich E, et al. Assimilation of GPS radio occultation data for an intense atmospheric river with the NCEP regional GSI system. Monthly Weather Review. 2011;**139**: 2170-2183. DOI: 10.1175/2011MWR3342.1

[43] Neiman PJ, Ralph FM, Wick GA, Kuo Y-H, Wee T-K, Ma Z, et al. Diagnosis of an intense atmospheric river impacting the Pacific Northwest: Storm summary and offshore vertical structure observed with COSMIC satellite retrievals. Monthly Weather Review. 2008;**136**:4398-4420

[44] Mears C, Wang J, Ho S-P, Zhang L, Zhou X. Global climate hydrological cycle, total column water vapor [in "State of the Climate in 2011"]. Bulletin of the American Meteorological Society. 2012;**93**(7):S25-S26. DOI: 10.1175

[45] Mears C, Wang J, Smith D, Wentz FJ. Intercomparison of total precipitable water measurements made by satellite-borne microwave radiometers and ground-based GPS instruments. Journal of Geophysical Research – Atmospheres. 2015;**120**:2492-2504. DOI: 10.1002/2014JD022694

[46] Mears C, Ho S-P, Peng L, Wang J. Total column water vapor, [in "State of the Climate in 2016"]. Bulletin of the American Meteorological Society. 2017; **98**(8):S93-S98. DOI: 10.1175/2017BAMSStateoftheClimate.1

[47] Strow LL, Hannon SE, DeSouza-Mackado S, et al. An overview of the AIRS radiative transfer model. IEEE Transactions on Geoscience and Remote Sensing. 2003;**41**(2):303-313

[48] Liu C-Y, Li J, Ho S-P, Liu G-R, Lin T-H, Young C-C. Retrieval of atmospheric thermodynamic state from synergistic use of radio occultation and hyperspectral infrared radiances observations. IEEE Journal of Selected Topics in Applied Earth Observations and Remote Sensing. 2016;**9**:744-756. DOI: 10.1109/JSTARS.2015.2444274

[49] Wick GA, Kuo Y-H, Ralph FM, Wee T-K, Neiman PJ, Ma Z. Intercomparison of integrated water vapor retrievals from SSM/I and COSMIC. Geophysical Research Letters. 2008;**28**:3263-3266